Tobias Baas

ECG Based Analysis of the Ventricular Repolarisation in the Human Heart

Vol. 18
Karlsruhe Transactions on Biomedical Engineering

Editor:

Karlsruhe Institute of Technology
Institute of Biomedical Engineering

Eine Übersicht über alle bisher in dieser Schriftenreihe erschienenen Bände
finden Sie am Ende des Buchs.

ECG Based Analysis of the Ventricular Repolarisation in the Human Heart

by
Tobias Baas

Dissertation, Karlsruher Institut für Technologie
Fakultät für Elektrotechnik und Informationstechnik, 2012

Impressum

Karlsruher Institut für Technologie (KIT)
KIT Scientific Publishing
Straße am Forum 2
D-76131 Karlsruhe
www.ksp.kit.edu

KIT – Universität des Landes Baden-Württemberg und nationales
Forschungszentrum in der Helmholtz-Gemeinschaft

KIT Scientific Publishing 2012
Print on Demand

ISSN: 1864-5933
ISBN: 978-3-86644-882-7

ECG Based Analysis of the Ventricular Repolarisation in the Human Heart

Zur Erlangung des akademischen Grades eines

DOKTOR-INGENIEURS

an der Fakultät für

Elektrotechnik und Informationstechnik

des Karlsruher Instituts für Technologie (KIT)

genehmigte

DISSERTATION

von

Dipl.-Ing. Tobias Baas

geboren in:

Worms am Rhein

Tag der mündlichen Prüfung: 09. Juli 2012
Referent: Prof. Dr. rer. nat. Olaf Dössel
Korreferent: Prof. Dr.-Ing. Hartmut Dickhaus

Acknowledgement

Diese Arbeit ist zwischen Juli 2008 und Juni 2012 am Institut für Biomedizinische Technik des Karlsruhe Institut für Technologie (KIT) entstanden. Am Gelingen dieser Arbeit haben viele Menschen einen Anteil, denen ich im folgenden danken möchte.

An erster Stelle danke ich Prof. Dr. rer. nat. Olaf Dössel, der es mir ermöglichte an diesem Thema zu arbeiten und die vorliegende Arbeit zu verfassen. Er stand mir während der gesamten Zeit beratend zur Seite und hat mich stets motiviert und gefördert. Hierfür gilt mein besonderer Dank.

Einen herzlichen Dank möchte ich Herrn Prof. Dr.-Ing. Hartmut Dickhaus für das Interesse an meiner Arbeit und die Übernahme des Korreferates aussprechen.

Danken möchte ich auch meinen Kollegen am Institut, für das stets angenehme Arbeitsklima, die vielen konstruktiven Diskussionen, die große Hilfsbereitschaft und die damit verbundene Unterstützung.

Für Ihre Unterstützung danke ich auch allen Studenten, die mit ihren wissenschaftlichen Arbeiten mit zum Gelingen dieser Arbeit beigetragen haben. Gustavo Lenis, Tobias Oesterlein und Martin Pfeifer möchte ich für ihr überdurchschnittliches Engagement an dieser Stelle besonders danken.

Ich danke meinen Eltern, Großeltern, meinem Bruder und allen Freunden die mich stets liebevoll unterstützten, motivierten und mir Halt gaben. Auch Sie haben einen großen Anteil zum Gelingen dieser Arbeit beigetragen.

Besonders danke ich meiner Freundin Cornelia für ihre Unterstützung, die große Geduld und die schönen Zeiten zusammen, die über manche Anstrengung hinweghalfen.

Karlsruhe im Juni 2012   *Tobias Baas*

"Erinnerungen verblassen und des Tages Ruhm vergeht,
die Spuren, die wir heute zieh'n, sind morgen schon verweht.
Doch in uns ist die Sehnsucht, dass etwas von uns bleibt,
ein Fußabdruck am Ufer, eh' der Strom uns weitertreibt."

Reinhard Mey

Contents

List of Abbreviations

A/D	Analog to Digital
AP	Action Potential
APD	Action Potential Duration
AR	AutoRegressive
BCL	Basic Cycle Length
BSAT	BioSignal Analysis Toolbox
CI	Coupling Interval
CMRR	Common Mode Rejection Ratio
CP	Compensatory Pause
CR	Correct Rate
DFT	Discrete Fourier Transform
DLR	Driven Right Leg
DWT	Discrete Wavelet Transform
EEG	ElectroEncephaloGraphy
EOB	End Of Block
FDA	Food and Drug Administration
FFT	Fast Fourier Transform
FT	Fourier Transform
GUI	Graphical User Interface

HRT	Heart Rate Turbulence
IBT	Institute of Biomedical Engineering
ICH	International Conference for Harmonization
iDFT	inverse Discrete Fourier Transform
iFFT	inverse Fast Fourier Transform
iFT	inverse Fourier Transform
iSWT	inverse Stationary Wavelet Transform
KIT	Karlsruhe Institute of Technology
LA	Left Arm
LL	Left Leg
Mox	Moxifloxacin
MVAR	MultiVariate AutoRegressive
Op-amp	OPerational AMPlifier
PAC	Premature Atrial Contraction
PCA	Principal Component Analysis
PPV	Positive Predicted Value
PVC	Premature Ventricular Contraction
RA	Right Arm
RBF	Radial Basis Function
RL	Right Leg
RMS	Root Mean Square
SA	SinoAtrial
SE	Sensitivity
SNR	Signal to Noise Ratio
SVM	Support Vector Machine
SWT	Stationary Wavelet Transform

TdP	Torsade de Points
THEW	Telemetric and Holter ECG Warehouse
tQT	thorough QT
TRF	Transfer Function
UnC	Unknown Compound
WCT	Wilson Central Terminal
WT	Wavelet Transform

List of Figures

List of Tables

1

Introduction

1.1 Motivation

Since the Dutch physician William Einthoven established the electrocardiograph in 1903 the analysis of the Electrocardiogram (ECG) has become the most important non invasive diagnostic method in cardiology. The ECG is a biosignal that provides diagnostic relevant information of the heart. However, the extraction of those information is challenging as the humans heart beats up to 100,000 times a day and often recordings of a day and more are necessary. Hence, long term ECG signals are common and inspection by the human eye is time consuming. Thus computer assisted methods are need. Every heart beat has to be detected and the boundaries of the ECG waveforms delineated. Heart beats have to be classified and time intervals such as e.g. RR- and QT interval need to be extracted to get the relevant information, which can primarily help to diagnose diseases. However, while the signal analysis is a fascinating and important domain, the patient and the physician are first of all interested in clinical information. In this thesis a combination on both is presented.

ECG recordings provide diagnostic relevant information on the de- and repolarisation sequences of the heart. A modification of these electrical processes is dangerous. To detect these changes long term ECG recordings are analysed. Modifications of the repolarisation sequence is in focus, as it is assumed to be pro-arrhythmic. In the 1990s, some non cardiac drugs e.g. Sotalol or Cisapride [1] have been associated with sudden cardiac death and were withdrawn from the market. The genesis of arrhythmia resulting into a dangerous ventricular tachyarrhythmia is often associated with a prolongation of the QT interval in the ECG. Thus, the *Food and Drug Administration* (FDA) which is responsible for drug approval in the USA, dictates the analysis of the QT interval in drug safety studies for the approval of all new drugs. As the analysis of long term ECG recordings with its huge amount of heart beats is only practical by computer algorithms, the FDA additionally demands for further research in the field of fully automatic T wave delineation.

However, the relationship between the prolongation of the QT interval and the arrhythmic risk is controversially discussed and so far not fully understood. Not every prolongation inevitably leads to a tachyarrhythmia. Moreover it is assumed that the change of the repolarisation pattern and not a delayed repolarisation is responsible for the genesis of tachyarrhytmias. Therefore, an analysis of the ventricular repolarisation is required.

The morphology of the T wave is the result of the transmembrane voltage gradients of the ventricles during the repolarisation. Changes in the repolarisation processes will lead to changes in the morphology of the T wave. The analysis of the T wave in the surface ECG is reported, but not exploited clinically.

1.2 Focus of the Thesis

New methods of ECG signal processing and subsequent T wave morphology analysis are presented. The methods are used to investigate the repolarisation sequence of the ventricles in the surface ECG e.g. in drug safety studies. Additionally the influence of heart rate and ectopic beats on the repolarisation of the ventricles is analysed and discussed. Recapitulatory, changes of the repolarisation related to medication, sporting activity and diseases were in focus.

To analyse the morphology of the T wave, an improved signal processing was necessary. An algorithm for fully automatic delineation of the QRS complex and the T wave has been developed. The delineation algorithm provides the waveform boundaries for every single heart beat which made a successive analysis of the T wave morphology of every beat possible. A beat classification was done to detect ectopic beats in the ECG. The whole signal processing workflow can be separated into the following parts:

- Fourier and Wavelet based ECG filtering
- Wavelet based QRS delineation
- Correlation based T wave delineation
- Beat classification based on a *Support Vector Machine* (SVM)

- Time series generation and wave extraction (whole heart beat, QRS complex and T wave)

All signal processing steps were embedded into a software named *BioSignal Analysis Toolbox* (BSAT), developed during this research project by the author.
The investigation of the ventricular repolarisation can be separated into the following five projects:

- Investigation of drug induced QT prolongation
- Analysis of the T wave morphology to investigate drug induced repolarisation changes
- Investigation of heart rate induced changes of the T wave morphology
- Model parameter estimation in MVAR[1] models to detect abnormalities in QT/RR coupling
- Influence of Premature Ventricular Contractions (PVC) on the repolarisation of subsequent heart beats

1.3 Structure of the Thesis

Part I covers basic medical, mathematical and technical principles:

- **Chapter 2** gives a short introduction into the medical fundamentals, together with the anatomy and physiology of the heart. Additionally, the cardiac electrophysiology and the ECG are introduced. In the last section different ECG lead systems are in focus.
- **Chapter 3** Introduces the mathematical basics. Contents are: Fourier transform, Wavelet transform, Principal Component Analysis (PCA), Hermite basis functions, an introduction to the support vector machine and some remarks on statistical tests used in this work.
- **Chapter 4** is focussed on the ECG recording technique. The basics of ECG data acquisition are introduced. In addition an ECG recording device named *BSAT-1012*, developed by the author during this thesis, is introduced in this section.
- **Chapter 5** presents the ECG data used in this thesis.
- **Chapter 6** gives an overview over the state of the art in ECG signal processing.

Part II presents utilised signal processing methods:

---

[1] *MultiVariate AutoRegressive* (MVAR)

- **Chapter 7** Describes the developed signal processing algorithms. It starts with the filtering of the ECG. Next the QRS detection is in focus. The delineation of the T wave is followed. Parts of the T wave delineation algorithm have been developed during the supervised Diploma Thesis of Franz Gravenhorst [2]. The algorithm has been presented on a scientific conference [3].

  The beat classification is introduced next. Parts were created in a supervised Student Research Project by Gustavo Lenis [4].

  The generation of the time series is described in the following section. At the end of Chapter 7 a short introduction on the developed software toolbox BSAT can be found.

Part III presents the investigations and results of the data analysis for the different projects:

- **Chapter 8** is focussed on the ventricular repolarisation. In the first section an analysis of drug (Moxifloxacin) induced QT prolongation is introduced. Parts have been presented on the scientific conference [5].

  In the second section changes of the T wave morphology due to the drug Moxifloxacin and an unknown compound are presented. Some of the corresponding results have been published on a scientific conference [6]. Also a supervised Diploma Thesis was undertaken by Ksenja Gräfe in the field of this research project [7].

  The influence of the heart rate on the previously introduced morphology based parameters is presented in the last section of Chapter 8. Parts of the results shown were created during the supervised Student Research Project of Martin Pfeifer [8].

- **Chapter 9** presents the estimation and analysis of MVAR model parameter to detect changes in the QT/RR coupling. Parts of this project have been created during the supervised Diploma Thesis of Heidrun Köhler [9] and the supervised Student Research Project of Tobias Oesterlein [10, 11].

- **Chapter 10** presents results of the heart rate turbulence analysis and the effects of ventricular ectopic beats on the repolarisation sequence of subsequent heart beats. Parts of the results have been created during the supervised Diploma Thesis of Gustavo Lenis [12].

Part I

Basic Principles

2

Medical Background

In this chapter a short introduction into the anatomy and physiology of the humans heart is given. Within this introduction, fundamental information on the anatomy and physiology of the heart is presented, while the often extremely complex underlying processes are not discussed. For detailed information the author refers to specialised literature like e.g. [13, 14, 15, 16].

2.1 Cardiac Anatomy

The human heart is a hollow muscular organ located near the anterior chest wall, directly posterior to the sternum. In adult humans it measures typically around 12 cm from the base to the apex [13]. The heart pumps blood continuously into the cardiovascular system to distribute oxygen and remove metabolic waste. The cardiovascular system can be divided into the pulmonary circuit and the systemic circuit. In the pulmonary circuit the blood is pumped to the gas exchanging surfaces of the lungs, while in the systemic circuit the blood is pumped throughout the whole body. In a healthy adult who rests, the heart pumps approximately 5.6 litres of blood per minute. The whole blood volume of an adult human is about 4-5 litres (about 7 % of the fat free body mass) [17]. Figure 2.1 shows a schematic illustration of the human heart. It comprises four chambers: Two atria, and two ventricles. The thickness of the heart muscle depends on the load in the respective chamber. For example, the wall of the right ventricle is much smaller than the wall of the left ventricle. The left ventricle needs a higher strength, as it has to pump the blood through the systemic circuit with all organs. The blood pressure in this loop is about 80 to 120 mmHg, while the blood pressure in the pulmonary circuit is about 30 mmHg.

To guarantee unidirectional blood flow, a number of different valves exist inside the heart: One between the atrium and ventricle (right/left), one at the exit to the pulmonary arteries and one at the aorta.

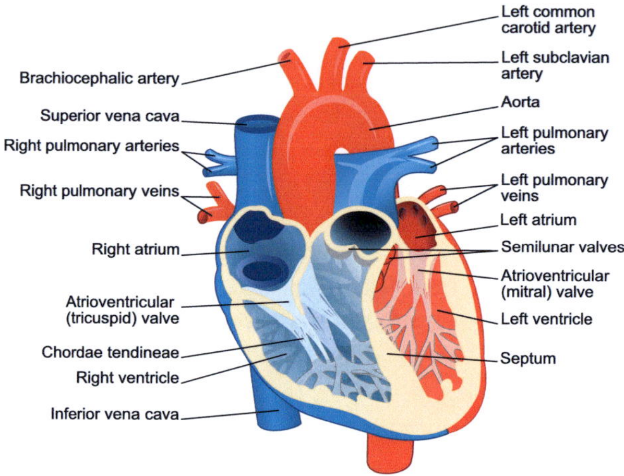

Fig. 2.1. Human heart: anterior view, coronal section. Figure adopted from [18]

2.2 Cardiac Electrophysiology

As any human cell, heart cells (myocytes) are encompassed by a cell membrane. This membrane separates the intra- from the extracellular space. Different ionic concentrations between intra- and extracellular space lead to a potential difference across the cell membrane. This is the so-called transmembrane voltage. The resting transmembrane voltage is about -90 mV. Myocytes are electrically excitable if the transmembrane voltage is raised from the resting voltage to more positive values. Different ionic channels, pumps and exchanges in the cell membrane are responsible for a controlled electrical excitation of the myocytes. This process can be divided into four phases. Figure 2.2(a) shows the conductivity of the three most important ionic channels. Figure 2.2(b) shows the Action Potential (AP). The four phases of the AP are now explained in more detail:

- **Phase 1**: An AP is triggered by rising the transmembrane voltage. Conductivity of the fast sodium channels increases and positively charged sodium flows into the cell. The transmembrane voltage increases further, as a positive feedback results. The rapid increase of the transmembrane voltage leads to the opening of all sodium channels and the transmembrane voltage reaches its peak at approximately 20 mV. The first phase ends when the sodium channels close after a few milliseconds.

- **Phase 2**: This is the so-called plateau phase. The transmembrane voltage remains more or less constant. The calcium conductivity is increased, which leads to a flow of calcium into the cell. This triggers a release of internally stored calcium and initiates the mechanical contraction [19].

- **Phase 3**: The voltage gated potassium channels open. The positively charged potassium ions inside the cell flow to the extracellular space. The transmembrane voltage returns to the resting potential.

- **Phase 4**: Potassium channels are open, while sodium and calcium channels are closed.

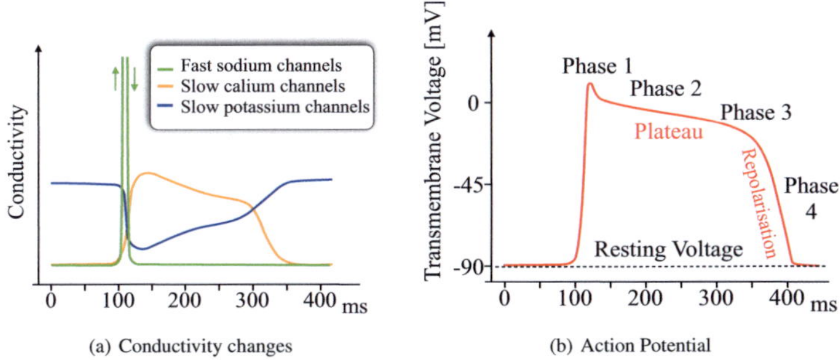

Fig. 2.2. Action potential of a myocyte (b) and corresponding ionic channel conductivity of sodium, calcium and potassium (a). Figure modified based on [20].

The so-called refractoriness prevents the myocytes from getting depolarized before the previous AP is nearly finished. Therefore the sodium channels cannot be reactivated until phase 4. This is an important safety feature to prevent cardiac arrhythmias.

The myocytes are connected through gap junctions to provide a continuous and directed excitation process leading to an optimal mechanical contraction. A trigger impulse from the sinus node thus leads to a successive excitation of all myocytes in the heart. The forwarding of the excitation is performed by specialised cells that ensure a fast and coordinated excitation of every region in the heart. Figure 2.3 illustrates this so-called electrical conduction system. The excitation then travels through the atrium to the SinoAtrial (SA) node. Similar to the sinus node, the SA node also has pacemaker capabilities, but with a lower frequency. If the trigger impulse from the sinus node is missing e.g. because of sinus node disease, the AV node fulfils the task of pacing the heart. From the AV node the excitation travels through the bundle of His to the bundle brunch into the Purkinje fibres.

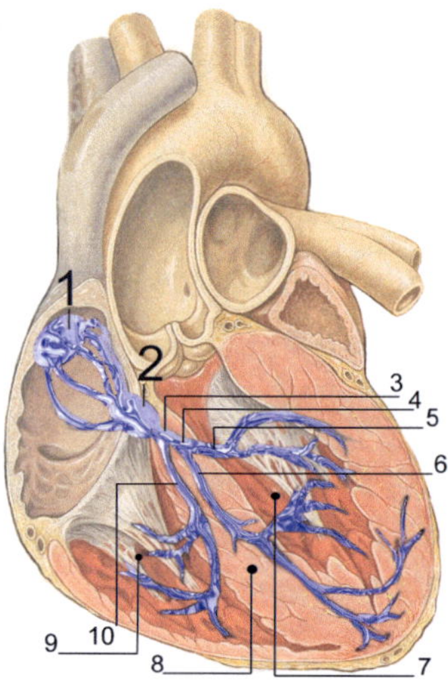

Fig. 2.3. Electrical conduction system of the heart. 1: sinoatrial node, 2: atrioventricular node, 3: bundle of His, 4: left bundle branch, 5: left posterior fascicle, 6: left-anterior fascicle, 7: left ventricle, 8: ventricular septum, 9: right ventricle, 10: right bundle branch. Figure adapted from [21].

2.3 The Electrocardiogram

During the excitation and repolarisation process of the heart, differences in the electrical potential distribution exist. The projection of this transmembrane voltage gradient can be measured at the body surface. This measurable biosignal is named ECG. Figure 2.4 shows a standard ECG with its different waveforms. The waves and segments in the ECG signal can be distinguished into the following parts:

- **P wave**: The excitation of the myocytes in the atrium leads to the P wave in the ECG. The P wave has a typical duration of about 50 to 100 ms [22].

- **PQ segment**: The baseline signal between P offset and QRS onset comes from the delay of the AV node. Normal duration of the PQ interval is about 120 to 200 ms [22][1].

- **QRS complex**: The depolarisation of the ventricles is referred to as QRS complex. The normal duration of a QRS complex is 60 to 100 ms [22].

- **ST segment**: The ST segment begins with the S peak (R peak if no S peaks exists) and ends with the beginning of the T wave. ST segment elevation or depression can be an indicator for ischemia or myocardial infarction.

- **T wave**: This wave comes from the transmembrane voltage gradients during the repolarisation process in the ventricles. The shape of the T wave provides information on the repolarisation sequence and can be used as a marker for some heart diseases [22].

- **U wave**: In some cases the U wave exists in the ECG. The origin of the U wave is not fully understood and controversially discussed in the community [24, 25, 26].

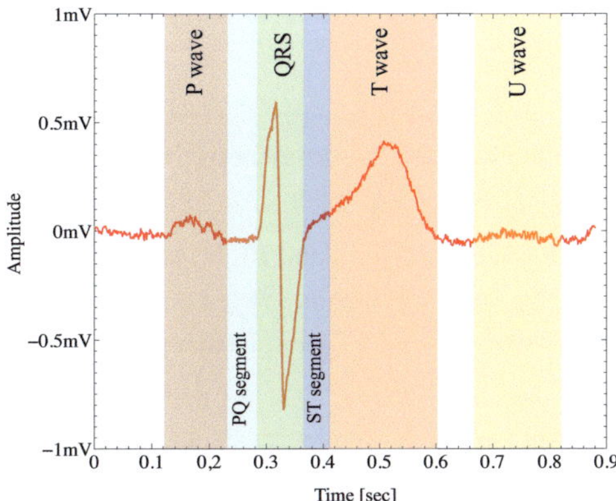

Fig. 2.4. Typical ECG signal with marked waves and segments. Signal recorded with BSAT 1012 ECG recorder.

---

[1] In [23], 46,129 healthy subjects have been analysed. RP interval turned out to be between 113 and 212 ms. Additional empirical values of the ECG segments can be found there.

2.3.1 ECG Lead System

Electrodes are used to measure the ECG signal at the body surface. The position of these electrodes is partly responsible for the resulting waveforms of the ECG signal. Depending on the kind of information that is needed and the genesis of the underlying potential distribution, different electrode positions are necessary. In this section the most common electrode positions are introduced. Most of the ECG signals used in this work have been recorded with one of these electrode arrangements.

Unipolar and Bipolar Leads

ECG leads are categorised in unipolar and bipolar. Bipolar leads have a positive and a negative electrode. The measured signal is the potential difference between those two electrodes.

Of course also unipolar leads need to measure a potential difference between two electrodes, but here one electrode is situated at a general reference point. This reference point can be created by combining different leads, e.g. limb leads in Figure 2.6, or by using a point with large distance to the heart.

Limb Leads

Einthoven introduced the bipolar measurement of the ECG by using the so-called limb leads [27]. The electrodes are placed:

- **Lead I**: Between the Left Arm (LA) and Right Arm (RA).
- **Lead II**: Between the Left Leg (LL) and Right Arm (RA).
- **Lead III**: Between the Left Leg (LL) and Left Arm (LA).

Figure 2.5 illustrates the electrode positions on the human body. The polarity is standardized and can be seen in the figure.

$$I = \phi_{LA} - \phi_{RA} \tag{2.1}$$

$$II = \phi_{LL} - \phi_{RA} \tag{2.2}$$

$$III = \phi_{LL} - \phi_{LA} \tag{2.3}$$

The sum of all potentials measured by the Einthoven limb leads is zero:

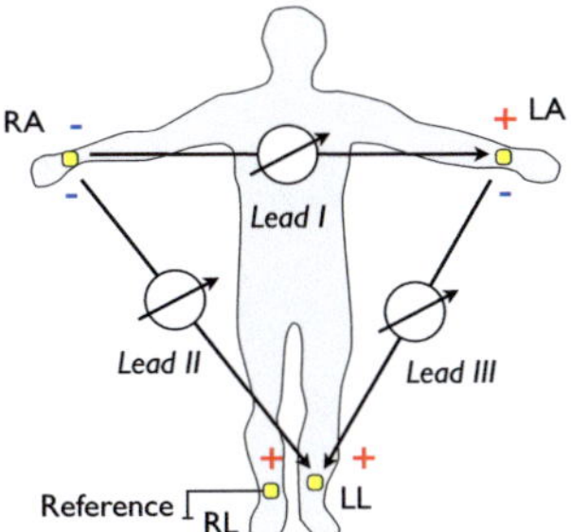

Fig. 2.5. Electrode position of limb leads (Einthoven's Triangle). Limb leads are measured bipolar.

$$0 = I + II + III \tag{2.4}$$

Augmented Limb Leads

Goldberger introduced the augmented limb lead electrode positions [28]. Similar to the electrode positions of the limb leads of Einthoven, electrodes are placed at the limbs. To increase the stability of the measurements, the two remaining electrodes are combined to form the reference point. Hence, a unipolar measurement results. The assembled resistors lead to an increase (augmentation) of the signal at the positive electrode. Figure 2.6 shows the electrode positions for the three augmented limb leads (*aVL*, *aVR* and *aVF*). In the labels of the leads, 'a' stands for the augmentation and 'V' for unipolar measurement.

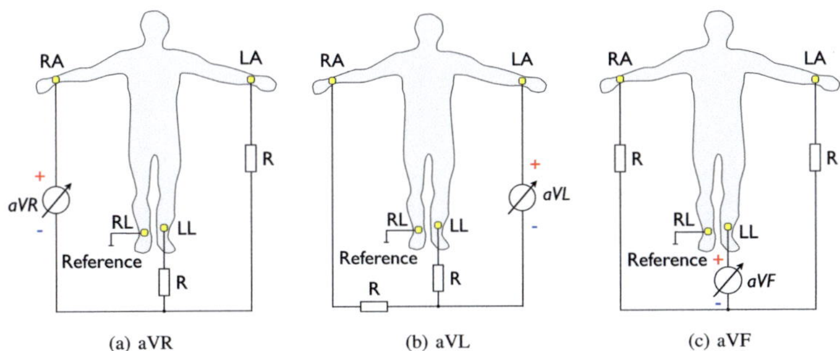

Fig. 2.6. Electrode position of augmented limb leads (Goldberger's leads). Augmented limb leads are measured unipolar. The resistors (*R*) are normally set to $5\,k\Omega$. In modern ECG devices the augmented limp leads are calculated without using any resistors.

In modern ECG devices the augmented limb leads are calculated. A realization with resistors is not done.

$$aVR = \phi_{RA} - \frac{\phi_{LA} + \phi_{LL}}{2} \tag{2.5}$$

$$aVL = \phi_{LA} - \frac{\phi_{RA} + \phi_{LL}}{2} \tag{2.6}$$

$$aVF = \phi_{LL} - \frac{\phi_{RA} + \phi_{LA}}{2} \tag{2.7}$$

Precordial Leads

Using the electrode position of the precordial leads, the potential at the positive electrode is measured against a common reference point which is the so-called *Wilson Central Terminal (WCT)* [29]. The *WCT* is generated by a combination of the three limb leads. Figure 2.7 shows the configuration. *WCT* can be calculated as follows:

$$\phi_{WCT} = \frac{\phi_{RA} + \phi_{LA} + \phi_{LL}}{3}. \tag{2.8}$$

The precordial leads measure the ECG signal at the chest. These electrodes are placed quite near to the heart. Due to this fact, the signals represent the potentials on the surface of the heart very well. However, not all regions of the heart surface can be measured by these electrodes, due to the shape of the body. In Figure 2.7 the six standard precordial leads are shown.

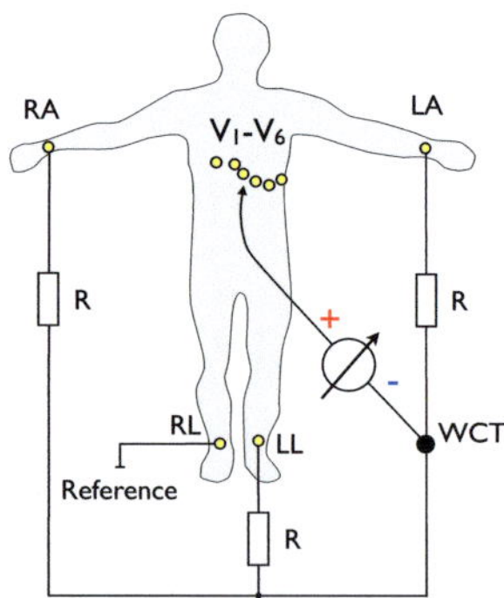

Fig. 2.7. Electrode arrangement of the precordial leads and the Wilson's central terminal (WCT). The resistors (R) are normally set to $5\,k\Omega$. In modern ECG devices the WCT is calculated without using any resistors.

The precordial leads can additionally be extended towards the right and to the back of the human body. For additional information on the lead systems, the vector cardiogram and the heart axes the author refers to [30, 16]. Additional Information on the medical interpretation of abnormalities in the different ECG leads can be found in [22].

3

Mathematical Basics

In this chapter the foundations of the mathematical methods used in this thesis are explained. The introduction is kept short and focussed strictly on the most important methods used in this thesis. For more detailed information and a derivation of the theory the author kindly refers to the literature. Corresponding references are provided in the respective sections.

3.1 Fourier Transform

The Fourier Transform (FT) is an integral transform. It transforms a time dependent function $y(t)$ into a frequency dependent function $Y(f)$. Hence, a representation in the time domain is transferred into a representation in the frequency domain. Information on the frequencies and the corresponding signal energy can be gained by the analysis of the signal in the frequency domain. The FT of a time dependent signal is called the spectrum of the signal.

3.1.1 The Continuous Fourier Transformation

The continuous FT of a continuous signal $x(t)$ is given by:

$$F\{x(t)\} = X(f) = \int_{-\infty}^{\infty} x(t) \cdot e^{-j2\pi ft} dt. \tag{3.1}$$

The corresponding Inverse Fourier Transform (IFT) is given by:

$$F^{-1}\{X(f)\} = x(t) = \int_{-\infty}^{\infty} X(f) \cdot e^{j2\pi ft} df \tag{3.2}$$

3.1.2 The Discrete Fourier Transform

As continuous time signals cannot be analysed, using computer based signal processing, the signals need to be transformed into discrete time signals. Let a continuous time signal $x(t)$ be sampled by a sample frequency $f_a = \frac{1}{t_a}$. The resulting signal $\widetilde{x}(t)$ results from a multiplication with a impulse train [31]:

$$\widetilde{x}(t) = x(t) \sum_{-\infty}^{\infty} \delta(t - n t_a) \tag{3.3}$$

$$x_n = x(n t_a) \, , \, n \in \mathbb{Z}. \tag{3.4}$$

Also the FT needs to be translated to be able to handle discrete time signals. The Discrete Fourier Transform (DFT) was developed for this purpose. A finite sequence of a signal $x(n)$ can be Fourier transformed by the DFT in the following way:

$$DFT\{x(n)\} = X(k) = \sum_{n=0}^{N-1} x(n) \cdot e^{-j2\pi \frac{kn}{N}} \tag{3.5}$$

$X(k)$ is the Fourier transform of the signal $x(n)$ with length N. The corresponding inverse DFT is defined by:

$$DFT^{-1}\{X(k)\} = x(n) = \frac{1}{N} \sum_{k=0}^{N-1} X(k) \cdot e^{j2\pi \frac{kn}{N}} \tag{3.6}$$

For an optimized computation of the DFT, the so-called Fast Fourier Transform FFT was introduced in 1967 by Brigham and Morrow [32]. The results of the FFT are equal to those of the DFT, while the number of operations needed to compute the result is significantly smaller. This is reached by the consideration of symmetries. A detailed introduction to the FT, DFT and FFT can be found in [31].

3.2 Wavelet Transform

The Wavelet Transform (WT) is a very common tool in the area of signal processing and image analysis. It is a integral transform with variable kernel. The kernel is the so-called wavelet. The signal under investigation is analysed by the WT to get information on both, time and frequencies. The introduction into this section is based on [33]. More detailed information can be found there.

3.2.1 The Continuous Wavelet Transform

The Wavelet transform is the projection of the signal under investigation $x(t)$ to the scaled and time shifted wavelet $\Psi_{a,b}(t)$ that is used for the analysis:

$$W_x^{\Psi}(a,b) = \langle x(t), \Psi_{a,b}(t) \rangle_t, \tag{3.7}$$

$$= \frac{1}{\sqrt{|a|}} \int_{-\infty}^{\infty} x(t) \Psi^* \left(\frac{t-b}{a} \right) dt. \tag{3.8}$$

In this equation, Ψ^* is the complex conjugate of Ψ, which is indicated by the asterisk. The factor $\frac{1}{\sqrt{a}}$ normalises the energy of the wavelet for all scaling factors a and shifts b. Only energy signals which fulfil the following rules can be used as wavelets:

- The signal energy needs to be finite:

$$\int_{-\infty}^{\infty} |\Psi(t)|^2 < \infty \tag{3.9}$$

- The Fourier transform $\Psi(f)$ is not allowed to have a zero component (admissibility condition):

$$C_{\Psi} = \int_{-\infty}^{\infty} \frac{|\Psi(f)|^2}{f} df < \infty \tag{3.10}$$

3.2.2 The Discrete Wavelet Transform

As already mentioned in Section 3.1, in computer based signal processing continuous time signals do not exist. Similar to the FT the WT has to be transformed into a discrete time version. A representation of the Wavelet coefficients can be calculated by [33]:

$$W_x^{\Psi}(m,k) = \langle x(n), \Psi_{m,k}(n) \rangle_n, \tag{3.11}$$

$$= \sum_{n=0}^{N-1} x(n) \cdot 2^{-\frac{k}{2}} \Psi^* (2^{-k}n - m) \tag{3.12}$$

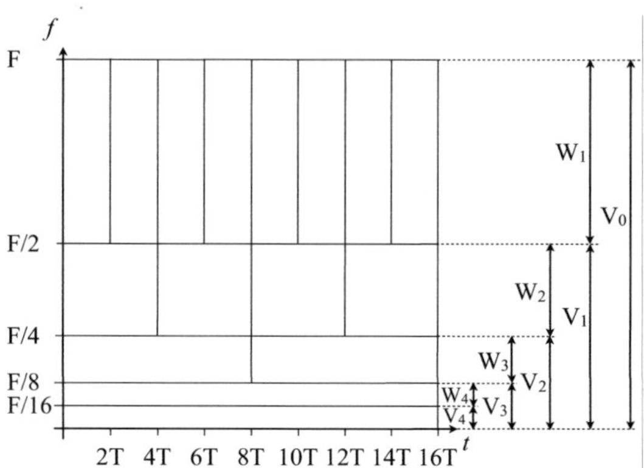

Fig. 3.1. Time-Frequency plane from the dyadic grid used in the DWT. Figure modified based on [33].

The scaling parameter a and the shifting parameter b in Equation 3.8 are discretised to m and k by a dyadic grid. Figure 3.1 shows an illustration of the time frequency plane that is based on the dyadic grid.

The Wavelet coefficients $W_x^{\Psi}(m,k)$ can be calculated as the projection of the discrete time signal $x(n)$ to the discrete time Wavelet functions $\Psi_{m,k}(n)$. However, due to the decreased computational demand, the method of multirate resolution filter banks is normally used to compute the Wavelet coefficients.

Multirate Resolution Filter Banks

The technique of calculating the Wavelet coefficients of a discrete time signal is the so-called Discrete Wavelet Transform (DWT). The decomposition of the signal is done in a recursive manner. It can be shown, that the signal $x(n)$, with a maximal frequency of $F = 0.5 \cdot f_a$ (f_a is the sampling frequency) can be approximated by [33]:

$$x(n) \approx x_0(n) = \sum_m c_0(m) \cdot \varphi(n-m), \tag{3.13}$$

with the scaling functions φ spanning a space V_0 (see Figure 3.1). The coefficients $c_0(m)$ can be calculated by the projection of $x(t)$ onto $\varphi_{m,0}(t)$.

$$c_0(m) = \sum_{n=0}^{N-1} x(n) \cdot \varphi^*(-(m-n)) \tag{3.14}$$

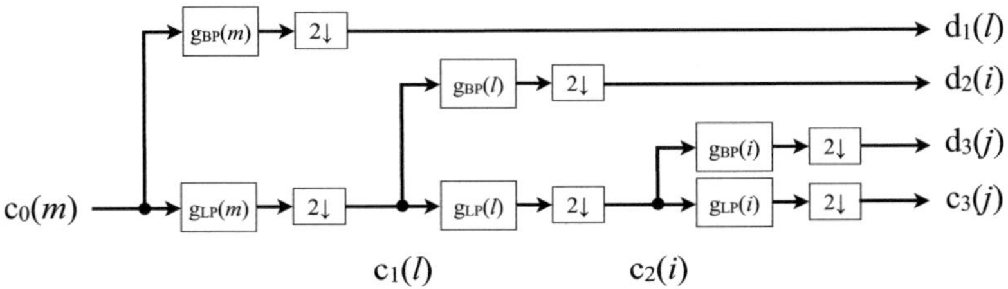

Fig. 3.2. Cascading and filter banks in the DWT. Figure adapted from [33]

As the Wavelets have a band pass characteristic, based on c_0 a recursive decomposition is made. The coefficients $c_k(m)$ are decomposed into the coefficients $c_{k+1}(l)$ and $d_{k+1}(l)$ by a convolution with the impulse response of a low pass filter $g_{LP}(m)$ and a band pass filter g_{BP} and a subsequent down sampling by 2:

$$c_{k+1}(l) = c_k(m) * g_{LP}(m) \,|_{m=2l} \tag{3.15}$$

$$d_{k+1}(l) = c_k(m) * g_{BP}(m) \,|_{m=2l} \;. \tag{3.16}$$

The scale- and wavelet functions are not used explicitly for this computation. By increasing k to $k+1$ the frequency resolution is increased by 2, while the time resolution is decreased by 2. Figure 3.1 illustrates the relationship for the different spaces W_k and V_k and the time and frequency relationship. The recursive computation of the coefficients c_{k+1} and d_{k+1} based on c_k, using low- and band pass, is illustrated in Figure 3.2. The coefficients $d_k(m)$ of the scaled subspaces W_k represent the Wavelet transforms:

$$W_x^{\Psi}(m,k) = d_k(m)\,. \tag{3.17}$$

Such a computation of the DWT is known as *Multirate Resolution Filter Bank*. A derivation of the scaling function and the discrete wavelet transform can be found in [33, 34]. For computation of the DWT Matlab has been used in this work .

3.3 Principal Component Analysis (PCA)

In 1933 Hotelling introduced the Principal Component Analysis (PCA)[1] as a method to transform a set of data $\boldsymbol{X} = (X_1 \ldots X_N)^T$ with respect to a new basis

---

[1] Alternative names are Karhunen Loève transform or empirical orthogonal functions.

with orthogonal linear projections [35, 36]. The projections are ordered by their variance. The PCA is often used as a dimensionality reduction method e.g. in data compression, pattern recognition or image analysis. A detailed description of the PCA and its different application field is given in [36]. In this section the author wants to introduce the PCA based on a typical ECG signal processing example. The example is adapted from [33, 37]. In [37] additional ECG signal processing techniques using the PCA can be found.

To detect outliers in an assembly of QRS complexes, which have been cut out of a continuous ECG signal, the PCA is used. All M signals of length N, each containing one QRS complex $(\boldsymbol{x}_m)$, are stored into a matrix:

$$\boldsymbol{X} = [\boldsymbol{x}_1 \cdots \boldsymbol{x}_M] = \begin{bmatrix} x_{1,1} & \cdots & x_{1,M} \\ \vdots & \ddots & \vdots \\ x_{N,1} & \cdots & x_{N,M} \end{bmatrix}, \tag{3.18}$$

$$\text{with } m := [1 \cdots M] \text{ and } n := [1 \cdots N]$$

The mean QRS signal is computed by:

$$\widehat{\boldsymbol{x}}(n) = \frac{1}{M} \sum_{m=1}^{M} \boldsymbol{x}_m(n) \tag{3.19}$$

and subtracted from all signals to calculate the average-free ECG vectors:

$$\boldsymbol{z}_m = [\boldsymbol{x}_m(1) - \widehat{\boldsymbol{x}} \cdots \boldsymbol{x}_m(N) - \widehat{\boldsymbol{x}}] . \tag{3.20}$$

These signals are combined to a matrix

$$\boldsymbol{Z} = [\boldsymbol{z}_1 \cdots \boldsymbol{z}_M]^T . \tag{3.21}$$

Out of the matrix $\boldsymbol{Z}$ the covariance matrix of the average free QRS signals is computed:

$$Cov\{\boldsymbol{Z}, \boldsymbol{Z}\} = \boldsymbol{C}_{\boldsymbol{ZZ}} = E\{(\boldsymbol{X} - E\{\boldsymbol{X}\})(\boldsymbol{X} - E\{\boldsymbol{X}\})^T\} \tag{3.22}$$

The covariance matrix can be decomposed to find the eigenvectors and the eigenvalues. The eigenvalue matrix $\boldsymbol{\Lambda}$ and eigenvector matrix $\boldsymbol{\Phi}$, resulting from the covariance matrix, are defined by:

$$C_{ZZ} = \boldsymbol{\Phi}\boldsymbol{\Lambda}\boldsymbol{\Phi}^T. \tag{3.23}$$

The eigenvalues (λ_n) and eigenvectors ($\boldsymbol{v}_m$) are paired and the vectors are sorted based on their eigenvalues.

$$\lambda_{n'} \leq \lambda_n, \text{ with } n' > n. \tag{3.24}$$

The scores $\boldsymbol{S}$ of the average free signals can then be calculated by:

$$\boldsymbol{S} = \boldsymbol{\Phi}^{T*} \cdot \boldsymbol{Z} = [\boldsymbol{s}_1 \cdots \boldsymbol{s}_M] = \begin{bmatrix} s_{1,1} & \cdots & s_{1,M} \\ \vdots & \ddots & \vdots \\ s_{N,1} & \cdots & s_{N,M} \end{bmatrix}, \tag{3.25}$$
$$\text{with } m := [1 \cdots M] \text{ and } n := [1 \cdots N]$$

Using the scores and eigenvectors, the transformed signals $\boldsymbol{x}_a$ can be reconstructed by:

$$\boldsymbol{x}_a = \widehat{\boldsymbol{x}} + \sum_{n=1}^{N} s_a(n) \cdot \boldsymbol{v}_n. \tag{3.26}$$

Equation 3.26 illustrates that the scores are multiplied with the eigenvectors. The sum in this equation represents the difference of the signal $\boldsymbol{x}_a$ to the mean signal $\widehat{\boldsymbol{x}}$.

For all signals with a good representation of the average QRS complex $\widehat{\boldsymbol{x}}$, the scores $s_{n,m}$ will decrease toward zero with increasing n. For outliers and QRS complexes with only small correlation to the average signal, the values of $s_{n,m}$ will not necessary decrease toward zero with increasing n. Using this fact, the Hotelling T^2 measure can be used to detect outliers. For all M signals the corresponding scores s_n are squared and divided by the corresponding eigenvalues:

$$T_m^2 = \sum_{n=1}^{N} \frac{s_m^2(n)}{\lambda_n}. \tag{3.27}$$

Outliers have increased values of T_m^2 compared to 'normal' QRS complexes. The decision whether the signal is a normal QRS complex or an outlier can be made by a predefined threshold. Outlier detection using the PCA and Hotelling T^2 are presented in Chapter 7 for several cases.

3.4 Hermite Basis Functions

Hermite basis functions can be used to describe an ECG waveform with only a few parameters. Differences in the morphology of a curve result in differences of the parameters of the Hermite basis functions [38]. In [39] an approximation of the QRS complex by the first three Hermite basis functions is introduced.
According to [39] a scaled version of an m-*th* order orthonormal Hermite function is defined by:

$$h_{m,\sigma}(t,\tau) = \frac{1}{\sqrt{2^n n! \sigma \sqrt{\pi}}} \cdot exp\left(-\frac{(t-\tau)^2}{2\sigma^2}\right) \cdot H_m\left(\frac{t-\tau}{2\sigma}\right). \qquad (3.28)$$

A Gaussian bell is present in this function. Hence, the waveform is approximated by several Gaussian functions resulting from multiplication with the Hermite polynomial H_m. The Gaussian bell has its centre at τ, while the variance is σ. Equation 3.29 represents the calculation of the Hermite polynomial of order m.

$$H_m(t) = (-1)^m \cdot exp(t^2) \cdot \frac{d^m}{dt^m}\left(exp(t^2)\right) \qquad (3.29)$$

The first three Hermite polynomials are given by:

$$H_0(t) = 1 \qquad (3.30)$$
$$H_1(t) = 2t \qquad (3.31)$$
$$H_2(t) = t^2 - 1. \qquad (3.32)$$

To approximate a waveform of time discrete values $x(n)$, a linear combination of Hermite functions can be used:

$$x(n) = \sum_{m=0}^{N} a_m \cdot h_{m,\sigma}(n,\tau). \qquad (3.33)$$

The parameters of the Hermite functions a_m are optimized for a best possible fit with the corresponding waveform.

3.5 Support Vector Machine

A support vector machine (SVM) is a classifier which predicts classes of a given observation. The SVM aims at increasing the distance between the borders of the classes. It maximises the margin, hence it is a so-called *large margin classifier*. This introduction to the SVM is a short overflew over the content presented in [36]. Detailed information on the SVM can be found there.

The SVM is a supervised learning method, generating a linear classifier which minimizes the risk of misclassification. A training data set is necessary:

$$\mathcal{L} = \{(\boldsymbol{x}_i, y_i) : i = 1, 2, \cdots, n\},\tag{3.34}$$

where $\boldsymbol{x}_i \in \mathbb{R}^p$, $p \in \mathbb{N}$ and $y_i \in \{-1, +1\}$. $\boldsymbol{x}_i$ is the feature vector and y_i represents the two possible classes. A separating function $f(x)$ classifies each point into one of the two classes:

$$C(x) = sign(f(x))\tag{3.35}$$

$$f(x) = \begin{cases} +1, & \text{if } \langle \boldsymbol{x}, \boldsymbol{\beta} \rangle + \beta_0 \geq 0 \\ -1, & else \end{cases}\tag{3.36}$$

where $\boldsymbol{\beta}$ is the *weight vector* of the classifier and β_0 is the *bias*. The term $\langle \boldsymbol{x}, \boldsymbol{\beta} \rangle + \beta_0$ is the decision boundary. It is a hyperplane in multi-dimensional cases. The weight vector $\boldsymbol{\beta}$ is a normal vector of the hyperplane. Every point in the data space x has the distance

$$D = \frac{\langle \boldsymbol{x}, \boldsymbol{\beta} \rangle + \beta_0}{||\boldsymbol{\beta}||}\tag{3.37}$$

to the hyperplane.

The Linearly Separable Case

During the training phase of the SVM, an optimal hyperplane is searched. The advantage of the SVM is that it finds the separation with the biggest possible margin ρ between the classes (see Figure 3.3(a)). This minimizes the misclassification probability. The margin is given by:

$$\rho = \min_i \left\{ y_i \cdot \left(\frac{\langle \boldsymbol{x}_i, \boldsymbol{\beta} \rangle + \beta_0}{||\boldsymbol{\beta}||} \right) \right\}.\tag{3.38}$$

ρ is chosen to be:

$$\rho = \frac{1}{||\boldsymbol{\beta}||} \tag{3.39}$$

The optimization problem which has to be solved to find the optimal hyperplane is to minimize Expression 3.40 according to $\boldsymbol{\beta}$ and β_0.

$$\underset{\boldsymbol{\beta},\beta_0}{\text{minimize}} \ \frac{1}{2}||\boldsymbol{\beta}||^2 \tag{3.40}$$

$$\text{subject to} \ \ y_i \cdot (\langle \boldsymbol{x}_i, \boldsymbol{\beta} \rangle + \beta_0) \geq 1 \ \ \text{for all } i. \tag{3.41}$$

The Linearly Nonseparable Case

In real applications it is not likely that all classes can be separated without any overlap. Hence, a soft-margin solution is introduced. Every observation (x_i, y_i) gets a sluck variable ξ_i, see Figure 3.3(b). ξ is a measure for the overlapping error. It is zero for all observations which do overlap. By considering and minimizing the overlapping errors ξ_i, Expression 3.40 gets to:

$$\underset{\boldsymbol{\beta},\beta_0}{\text{minimize}} \ \frac{1}{2}||\boldsymbol{\beta}||^2 + C \sum_{i=1}^{n} \xi_i \tag{3.42}$$

$$\text{subject to} \ \ y_i \cdot (\langle \boldsymbol{x}_i, \boldsymbol{\beta} \rangle + b) \geq 1 - \xi_i \ \ \text{for all } i \text{ and } \xi_i \geq 0. \tag{3.43}$$

C in Expression 3.42 is a regularisation parameter which has to be set by the user. The parameter controls the width of the margin and balances the two terms in the minimizing function.

The Non-Linear SVM

The introduced classifier is limited to linearly separable classes. In real applications the classes are not necessary linearly separable. The implementation of non linear curves to separate those classes is difficult. Thus, non linear input patterns are transformed by a non linear transformation $\Phi(x)$ into a new space, the so-called feature space. In some cases, the non linear problem becomes linear after the transform and makes the use of the linear SVM possible again.

It can be shown, that it is not necessary to implement a transfer function Φ, but

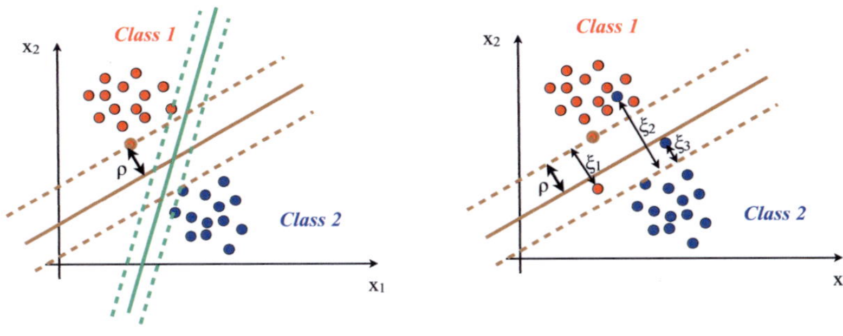

(a) Margin ρ of two possible separation lines.

(b) Separation plane with soft margin for the case of overlapping classes.

Fig. 3.3. Illustration of margin ρ and overlapping error ξ in the separation plane. Figure adapted based on [4].

rather a scalar function $K(x,y) = \langle \Phi(x)\Phi(y) \rangle$ which speeds up the computation. This function is called *Kernel Function*. In Figure 3.4 a transformation of the input data using polar coordinates is illustrated. In this work the Gaussian radial

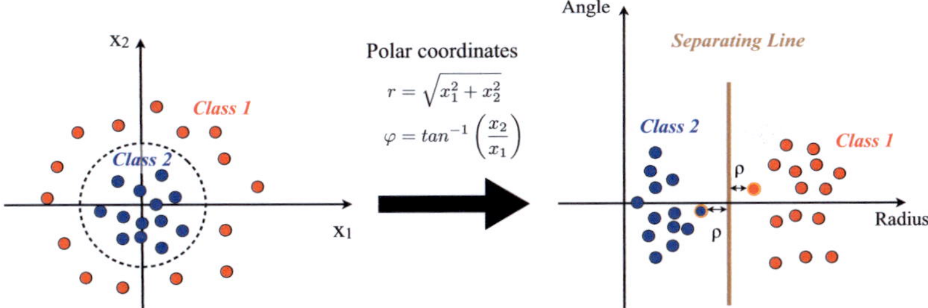

Fig. 3.4. Transformation of non linear data into a new space to become linear. The two classes on the left side can be separated by a circle. In the feature space after a transformation into polar coordinates the problem becomes linearly separable. Figure adapted based on [4]

basis function has been used. Detailed information on the SVM and the non linear implementation using kernel functions can be found in [36].

3.6 Statistical Hypothesis Test

Statistical tests are the common method to proof significance of results from clinical studies. Dependent on the preconditions and the assumption of the study, different statistic tests are available. The author of this work used three different statistic tests to proof the significance of changed descriptors. In this section a short mathematical introduction is given based on the content in [40]. Additional information on medical statistics can be found there.

3.6.1 Basics

All statistic tests used in this work return an information on the probability p of making a mistake by rejecting a hypothesis. The hypotheses to be proofed is named *null hypothesis H_0*. If H_0 is rejected, the *alternative hypothesis H_1* is accepted.

A significant level α is used to separate between a significant and a none significant p-value. Four α boundaries are commonly used in literature:

- $\alpha_{10\%}$ $(p < 0.1)$ lightly significant
- $\alpha_{5\%}$ $(p < 0.05)$ significance
- $\alpha_{1\%}$ $(p < 0.01)$ highly significant
- $\alpha_{0.1\%}$ $(p < 0.001)$ most significant

In this work a 'normal' significance with a significance level of 5% $(\alpha_{5\%})$ is used. This means all results are denoted as significant if a p-value of 0.05 or smaller is reached.

The p-value gives no information on the size of difference, the causal relationship, the reasons of the result or the clinical relevance of the results. To get the information on the size of the effect a confidence interval can be calculated.

In this thesis the groups of participants are constant within one study. Hence the assumption on paired samples can be made for all tests in this research work.

3.6.2 Sign Test

A sign test is one of the simplest statistical tests available. The test proofs for equal signs in both samples. The null hypothesis and the alternative hypothesis are given by:

$$H_0 : P(X < Y) = P(X > Y) = 0.5 \tag{3.44}$$

$$H_1 : P(X < Y) \neq P(X > Y) \neq 0.5. \tag{3.45}$$

Under the assumption of the null hypothesis the number of samples pairs with positive sign has to be equal to the number of samples pairs with negative signs.

Workaround:

Every sample pair gets a sign depending whether x_i or y_i is higher. If x_i and y_i are equal, the sample pairs are neglected. The number of positive and negative signs is counted and the proof value k is the smaller sum. The test result can be determined by the comparison of the critical values in the sign test table. The table can be found in [40].

Comparison with other statistical tests

The sign test has no special assumptions when evaluating clinical studies, as the observations are normally independent from each other. The question whether aligning the differences of x_i and y_i is meaningful, needs to be answered.

A sign test uses only very few information of the samples and thus the power of the test result is smaller compared to a Wilcoxon rank sum test or even a statistic t-test.

3.6.3 Wilcoxon Signum Rank Test

The Wilcoxon[2] rank sum test for paired samples with n observations compares the medians $\tilde{x}_i$ and $\tilde{y}_i$. The null hypothesis and the alternative hypothesis follows:

$$H_0 : \tilde{x}_i = \tilde{y}_i \tag{3.46}$$

$$H_1 : \tilde{x}_i \neq \tilde{y}_i . \tag{3.47}$$

The difference of every sample pair is calculated:

$$d_i = x_i - y_i . \tag{3.48}$$

Differences d_i with a value of zero are neglected. Values d_i are aligned for their size in increasing order to get a number r for its rank. The sum of all r-values of the positive d_i-values R_+ and negative d_i-values R_- are computed. The smaller number of R_+ and R_- is used as test value R.

---

[2] Frank Wilcoxon (1892-1965)

The corresponding p-value can be determined in a look up table for the test value R.

Comparison to other statistical tests

A precondition of this test is that the shape of the distribution of both samples have to be similar. If so, a symmetrical distribution of the differences d_i can be assumed. The power of a rank sum test is higher than a sign test, but lower than a t-test. The rank sum test does not use the shape of the distribution which makes it available for samples which are not normally distributed.

3.6.4 Student's *t*-Test

The paired sample t-test is the most powerful test used in this work. It analyses the statistical distribution of the samples. This test is often used to compare parameters affected during a therapy within the same population.

Workaround

The t-test analyses the position of the mean value. As the samples are paired, the difference is calculated:

$$d_i = x_i - y_i. \tag{3.49}$$

Out of these differences the mean is computed:

$$\hat{d} = \frac{1}{n} \sum_{i=1}^{n} d_i. \tag{3.50}$$

Additionally the variance is computed by:

$$S^2 = \frac{1}{n-1} \sum_{i=1}^{n} \left(d_i - \hat{d}\right)^2. \tag{3.51}$$

Using these values, a test value T can be calculated by:

$$T = \frac{\hat{d}}{S/\sqrt{n}}. \tag{3.52}$$

A table with percentage points of the t-distribution [41] is used to get the corresponding p-value out of T. The p-value again represents the probability of making a mistake when using the alternative hypothesis H_1.

Assumptions

The main assumption for a t-test is the normal distribution of the basic population. This is a claim difficult to proof, as the basic population is often unknown in available ECG studies. Non-normally distributed data alter the sensitivity of the test, if the null hypothesis is rejected [42]. Arguments pro and con normal distribution of the sample can be made looking at the histogram or calculating moments of the distribution.

In case of paired samples, normally symmetrically distributed differences d_i are satisfactory. This is not as difficult to proof as the claim of normal distribution of the basic population.

For small populations $n < 10$, it is recommended to use Wilcoxon signed rank test. For non symmetrical distribution of d_i sign tests are advantageous.

4

Technical Fundamentals of ECG Data Acquisition and Holter ECG

4.1 ECG Amplifier

In this section a short introduction into the recording of biosignals from the body surface is given. Assumptions are made for an ECG amplifier, however the same technique is used for other recordings from the body surface[1]. Contents of this section are adapted from [43]. Additional information on biosignal amplifiers can be found there.

To record a surface ECG, signals in a typical range between $1\,\mu V$ and $10\,mV$ and a frequency range from about $0.3\,Hz$ to $120\,Hz$ have to be recorded. These very low voltage signals have to be amplified to display or store the signal. Other signals on the human's body surface, like powerline interference or noise, are superposed. These signals often have 100 or 1000 times higher amplitudes which makes a recording of the biosignal like the ECG difficult. Hence, superimposed noise and interference signals have to be suppressed.

Using an instrumentation amplifier is the common way to record biosignals. This amplifier rejects noise and interference signals, by amplifying only differential signals. Noise and interference signals are more or less equal anywhere on the body surface. Measuring a potential difference between two electrodes on the body surface leads to very similar potentials in both leads for noise and interference signals, whereas the signals coming from the heart are highly different. Biosignals are different on the body surface and thus different in both leads. An instrumentation amplifier, boosts differences between its inputs and rejects common mode signals. The "quality" of an instrumentation amplifier is measured by the damping of common mode signals. The Common Mode Rejection Ratio (CMRR) is defined as the ratio between differential mode gain, G_D and common mode gain G_C.

---

[1] For example ElectroEncephaloGraphy (EEG): Amplifier settings, like amplification factor, adapted to the signals.

Figure 4.1 shows a schematic of the influence on the differential amplifier. The output voltage is a sum of both, the ECG signal (differential signal) V_{ECG} with its differential gain G_D and the common mode signal V_C with its common mode gain G_C. A further assumption has to be fulfilled: the amplifier has to have symmetrical inputs. Only if signalling paths of both inputs are equal, common mode signals are not amplified by the differential gain. Equation 4.1 shows the dependency of the output voltage on common mode and differential mode signals.

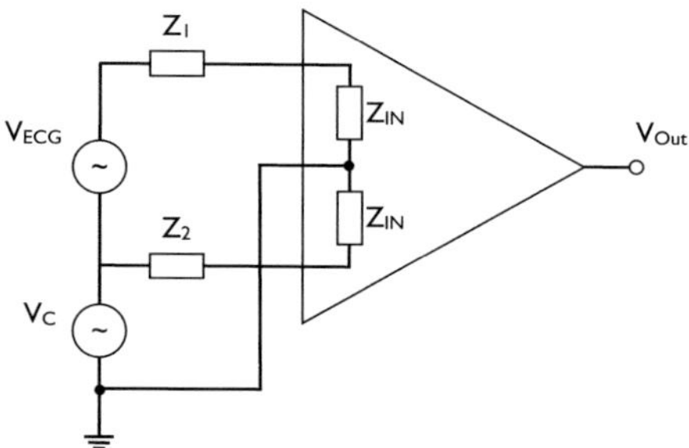

Fig. 4.1. Influence of common mode and differential mode signals on a differential amplifier

$$V_{Out} = G_D \cdot V_{ECG} + \frac{G_D V_C}{CMRR} + G_D \cdot \left(1 - \frac{Z_{in}}{Z_{in} + Z_1 + Z_2} \right) \tag{4.1}$$

Figure 4.2 shows a standard ECG circuit with adapted driven right leg circuit. The main ECG circuit represents an instrumentation amplifier. A short explanation on the instrumentation amplifier is given in the next section.

Instrumentation Amplifier

An instrumentation amplifier typically consists of a difference amplifier and an enhanced input circuit. The difference amplifier in Figure 4.2 consists of the operational amplifier (Op-amp) V_3 and the four resistors R_8 to R_{11}.

The expression corresponding to circuit in Figure 4.2 for the output voltage V_o with input voltages V_1 (at R_8) and V_2 (at R_9) is given by:

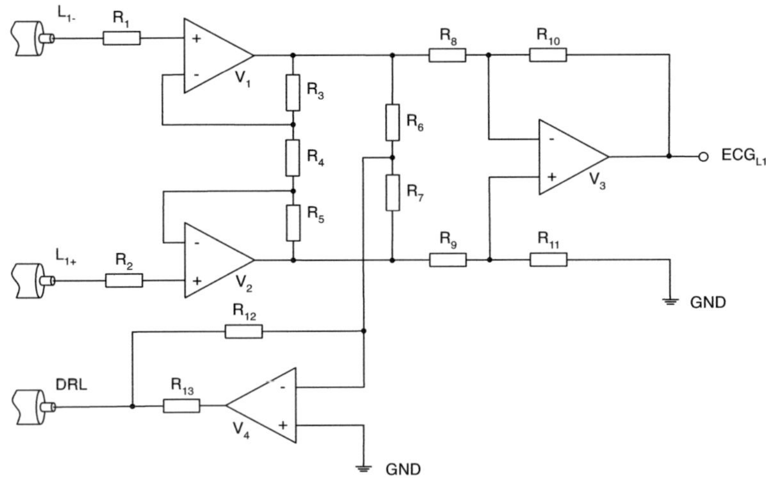

Fig. 4.2. Circuit diagram of a common ECG amplifier design

$$V_o = \frac{R_{11}(R_8 - R_{10})}{R_8(R_9 + R_{11})} V_2 - \frac{R_{10}}{R_8} \cdot V_1 \qquad (4.2)$$

with the assumption of equal resistor ratio:

$$\frac{R_8}{R_{10}} = \frac{R_9}{R_{11}} \qquad (4.3)$$

output voltage is given by:

$$V_o = \frac{R_{10}}{R_8} \cdot (V_2 - V_1) \qquad (4.4)$$

Further information on the difference amplifier in general can be found in [44]. The high accuracy of equal resistors as claimed in Equation 4.1 mainly determines the quality of the instrumentation amplifier.

Resistors R_8 and R_9 represent the input impedance of the difference amplifier. Without any precircuit these resistors limit the maximum possible input impedance of the amplifier. These resistors cannot be set to some $100\ M\Omega$. A high CMRR needs very high input impedances, as the source impedances are not equal and constant in both leads. A variation of some $1000\ \Omega$ is not unusual. To reach a

CMRR of over $100\,dB$, an input impedances of about $10^9\,\Omega$ is necessary. Inverting amplifiers are connected to the input path of the circuit. This helps to increase the input impedances and further amplifies the signals. The input circuit with both amplifiers has to be designed exactly symmetrical to prevent the differential gain from increasing common mode signals, as shown in Equation 4.2.

Driven Right Leg

To reduce common mode signals on the body surface, the reference electrode is not directly connected to ground. It is connected to an inverting amplifier, which inverts and feeds back the common-mode voltage to the body surface. This leads to decreased common mode signals and thus an increased signal quality of the biosignal after amplification. The method is named Driven Right Leg (DRL), as it replaces the ground electrode at the right leg.

4.2 ECG Amplifier Development *"BSAT-1012"*

In the clinical environment a wide spectrum on ECG devices can be found. However, these devices are sold commercially and have a static setup which cannot be modified. To investigate the morphology of the T wave, a high resolution ECG device with a sampling rate of at least 500 Hz was desired. Also the characteristics of the analogue filter of the device have to be known, to interpret the results correctly. Further more, the recorded ECG data have to be transferred to the analysing software BSAT, which was developed in this thesis. A solution satisfying all these facts was not found which led to the decision to develop an ECG recorder for this research project. The developed device *BSAT-1012* has a sampling frequency of 1 kHz, a resolution of 12-bit and records three ECG leads in parallel.

4.2.1 Analogue Circuit Development

To amplify the biosignals of the heart at the body surface, an instrumentation amplifier manufactured by *Analog Devices* is used. This integrated circuit is specially designed for medical instrumentation and needs only few additional electric components to amplify a surface ECG signal. The device has a very high CMRR of minimum 86 dB at DC and a minimum CMRR of 80 dB at 5 kHz for an amplification factor $G = 1$. The input currents remain typically below 300 pA because of

Fig. 4.3. Photography of the IBT ECG recorder *BSAT-1012*

the JFET inputs [45].

Using the AD8220 keeps the analogue part of the ECG device small. Only few capacitors and resistors in the input path of the instrumentation amplifier are used as a first low pass. Diodes in the input path limit the input voltage in case of static electrification of the body for example.

The gain of the instrumentation amplifier is set through external resistors to a value of 13.52 . To damp the baseline wander, subsequent the amplifier a high pass filter is installed to limit the frequency of the measured ECG to a frequencies above 0.3 Hz.

A second amplifier in the filter path is used to amplify the ECG up to 2 V. This variable, non inverting amplifier can be set to a gain between 1 and 101. The gain is adjustable by a variable resistor and thus an optimal adaptation of the signal amplitude to the range of the A/D converter is possible. Normally the gain values have been between 80 and 100. A first order anti aliasing filter at the input of the AD converter limits the signal frequency to a maximum of 500 Hz.

4.2.1.1 Circuit Diagram

Fig. 4.4 shows the total wiring diagram of the device. The upper left area of the diagram shows the digital circuit with the microcontroller board. The upper right area shows the battery power supply circuit. The analogue part can be found in the lower part of the wiring diagram. To amplify the ECG from the body surface the JFET instrumentation amplifier AD8220 from *Analog Devices* has been used. Due to its small size, high CMRR over a broad frequency range, rail-to-rail output, and JFET inputs, amplification of biosignals from the body skin is one of the main operating fields for this product as *Analoge Devices* declares. Each of the three leads has its own analogue amplifier cascade. In this description, the signal path of the first lead is explained. Lead two and three work in exactly the same way.

From the electrodes on the body surface the signals are lead through the connectors, R_1 and R_2 lying inside the connector, ahead to R_3 and R_4 at the input path to the instrumentation amplifier. The resistors R_1 to R_4 together with the capacitances C_1 to C_3 build up a first low pass filter for the input signals. Additionally, the resistors increase the security in case of a technical fault. The currents which might theoretically flow to the body surface are limited by these resistors.

The signals are amplified in the AD8220 with a gain of 13.52. The gain of the AD8220 can be set by the resistor between the two R_g inputs of the instrumentation amplifier. For this device, the gain is given by Equation 4.5

$$G = 1 + \frac{49.4k\Omega}{R_G} = 1 + \frac{49.4k\Omega}{(R_5 + R6)||R_7} \qquad (4.5)$$

with serial resistors R_5 and R_6 equal to $24\,k\Omega$ and $R_7 = 4.3\,k\Omega$, an amplification factor of 13.52 results. Instrumentation amplifiers as the AD8220 amplify the difference of the signals between the inputs $-IN$ and $+IN$. Common mode signals coming from the body are suppressed.

At the output of the *AD8220* a first order analogue high pass filter is installed by C_4 and R_8. The characteristic frequency is set to 0.3 Hz. The high pass filter is installed to reduce the baseline wander of the measured signal.

Ahead a second amplifier is installed. The second amplifier comes as a non-inverting operational amplifier circuit [44]. The amplification factor is given by the two resistors R_9 and R_{10}. As R_{10} is a variable resistor, the amplification factor can be changed. Equation 4.6 shows the relation:

$$G = \left(1 + \frac{R_9}{R_{10}} \right) \tag{4.6}$$

The amplification of the device is done in two steps. This is advantageous, as the baseline wander is reduced by the high pass behind the instrumentation amplifier before it is fully amplified. The signal at the A/D converter thus has a smaller voltage range which increases the resolution.

After the second amplification the signal is low pass filtered to prevent the digitized signal from aliasing artefacts. This is done by a passive first order low pass filter out of R_{11} and C_6. The cut-off frequency of the low pass filter is 500 Hz.

4.2.2 Power Management of the Device

The *BSAT-1012* ECG system is powered by 4 AA-Batteries. The batteries are pairwise driven in serial and power the positive and negative system voltage, to provide ± 5 V operating voltage. Voltage of the batteries has to be transformed by a step-up DC-DC converter. In *BSAT-1012* the integrated circuit MAX756 [46] was chosen to step-up the battery voltage. The device MAX756 accepts input voltages down to 0.7 V and provides an output voltage of 3.3 V or 5 V. In *BSAT-1012*, it works with a typical operation circuit setup recommended by MAXIM. In the ECG-system, two of the step-up DC-DC converters are placed and coupled in series. Figure 4.4 shows the circuit.

The currents of the positive and negative operating voltages are different, as the positive operating voltage source drives the digital parts of the circuit with the microcontroller and the amplifiers, while the negative operating source only drives the amplifiers. Thus the effective positive current is about $I_{pos} \approx 310\,mA$, while the negative current is about $I_{neg} \approx 8.5\,mA$. This means the recording duration of the device is limited by the batteries of the positive voltage source. Batteries with a capacity of 2800 mAh result in a theoretical recording duration of about 9 hours.

4.2.3 Signal Resolution

The conversion of the analogue signal at the input of the A/D converter requires a reference voltage. The device uses a 2 V reference voltage. Hence, the input signal has to be in a range between 0 V and 2 V. With a resolution of 12-bit, 4096 values can be distinguished. The smallest possible signal resolution is thus 0.488 mV. With a maximum total gain of $13.5 \cdot 100 = 1350$, the maximal possible resolution

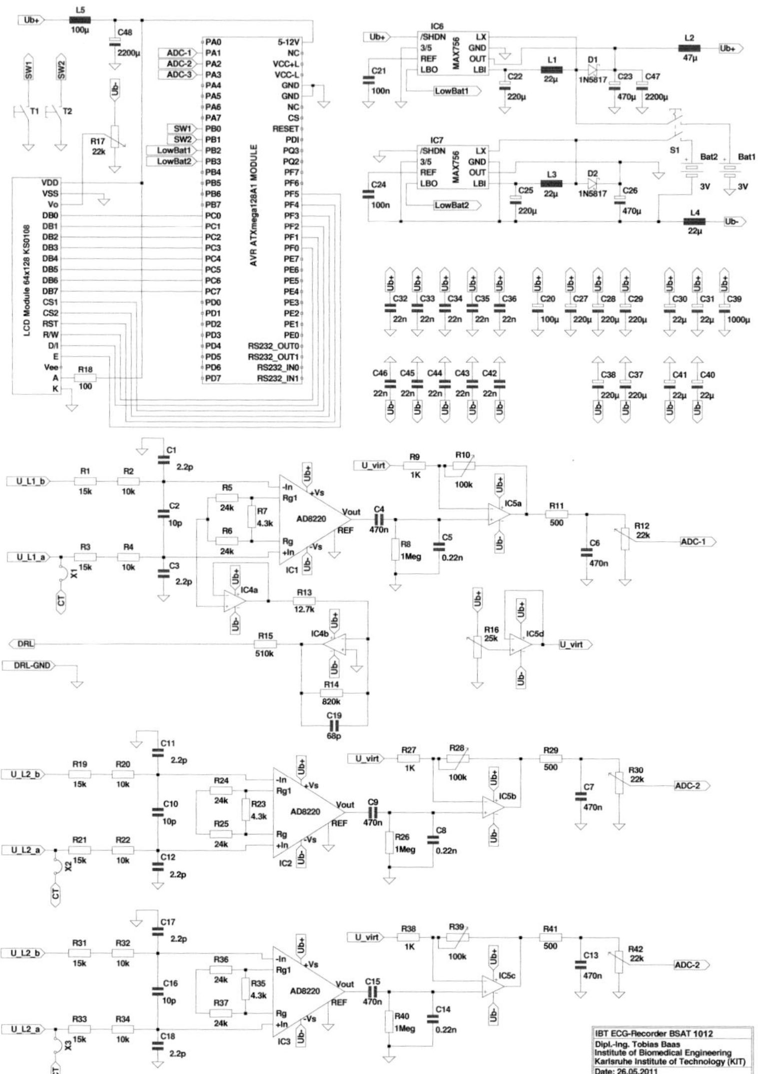

Fig. 4.4. Circuit diagram IBT ECG recorder *BSAT-1012*

Fig. 4.5. Photography of *BSAT-1012* inside

of the measured ECG signal is 362 nV.

4.2.3.1 Frequency Response Characteristic

To measure the frequency response characteristics, the second adjustable amplifier was set to an amplification factor of one. Thus, the amplitude of the input
signal could be higher and the Signal to Noise Ratio (SNR) of the input signal is
increased for the test. The frequency response was measured between 0.3 Hz and
5.5 kHz. The input signal was a sinus wave with an amplitude of 50 mV. The signal
was given to the connectors of the electrodes and the output was measured at the
connection to the microcontroller board (A/D converter input). Figure 4.6 shows
the result in a Bode diagram. It can be seen that the total analogue system of the
device has a bandpass characteristic. The characteristic frequency of the high pass
filter lies at about 0.3 Hz, while the characteristic frequency of the low pass filter
lies at about 500 Hz (each for the 3 dB point).

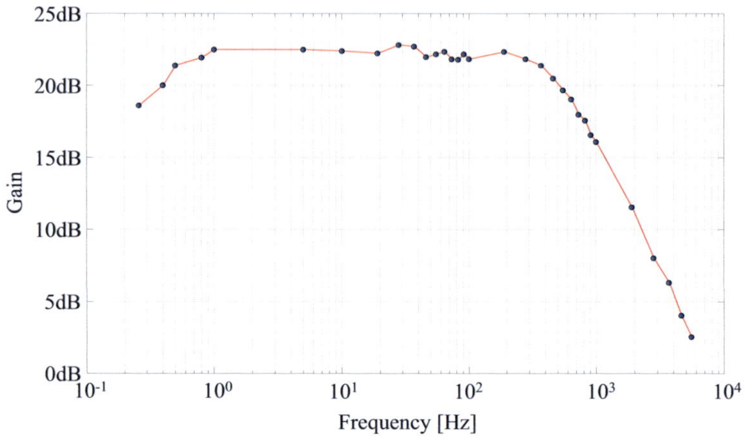

Fig. 4.6. Frequency Response Characteristic of *BSAT-1012*

4.2.4 Digital Signal Acquisition

4.2.4.1 Microcontroller

The digital signal processing of the ECG system *BSAT-1012* is done by the 8/16-bit microcontroller *ATXmega128A1* [47] manufactured by *Atmel*. The controller involves, besides other not used features, two eight channel 12-bit, 2 Msps A/D converters and eight 16-bit Timer/Counters. The controller has a 16 MHz clock and it is powered by 3.3 V. The *BSAT-1012* uses three channels of the first A/D converter to convert the analogue signal to numeric values with a resolution of 12-bit. The controller together with its peripheral is positioned on an evaluation board manufactured by ALVID [48]. The board includes the controller, a 3.3V power supply for the controller, a microSD-card reader and a 64 Mbit SDRAM which is not used by the device. The programming of the controller can be done either by a JTAG connection or an In-System-Programming (ISP) connector. The input power of 5V is transformed to 3.3V which is the operation voltage of the controller.

4.2.4.2 Firmware

The firmware of the device is written in the programming language *C*. Figure 4.7 shows a schematic of the firmware's structure. The device has two buttons *A* and

B to control the software. A 64x128 dot matrix display can show text and the recorded ECG signal of one lead.

Start up

When the device is started, the controller sets up the ports and the oscillator. Next the controller searches for the SD card. If there is no SD card in the microSD-card slot, the device displays a message and waits for a reset. If a microSD-card is detected the cards file system is searched for a file named *ECG.bin*. In case the *ECG.bin* file is not found the device shows a message and waits for a reset. If the file is found the device reads the header of the file and shows the previous recorded ECG-Files E_1 to E_N. The number of start- and end blocks is shown as well. After accepting the list by pressing the button *B* the device sets up a timer and the A/D converter which are needed to record a signal. To get a constant sampling frequency a 16-bit timer with interrupt execution is used. The timer is loaded by a numeric value which is counted down. When the timer reaches zero, an interrupt is executed. In the interrupt routine an A/D conversion of the three channels is done consecutively. The timers load value can be calculated by equation 4.7. Q_1 is the clock speed of the controller, L is the value to load to the timer, *PRE* is the pre-scale factor, set in the controllers register and T_{Int} is the time from the timers start to the interrupt execution.

$$L = \frac{T_{Int} \cdot Q_1}{PRE} \tag{4.7}$$

Measurement

To start the measurement the button *A* has to be pressed. The controller jumps into the main loop of the program. The timer is started and the system begins to record the signals. The data which is stored on a SD card has to fulfill the rules of the FAT-16 system. Thus, the controller internally saves the signal data until one block of 512 byte is recorded and writes the whole block to the SD card.

During main loop the timers interrupt routine, executing in equidistant time steps, converts and stores the voltage at the input of the A/D converter. The system is programmed with an oversampling and decimation. This was done to increase the accuracy of the measured signal. The oversampling and dividing results in a low pass filter of the signal which reduces the noise. Hence, the anti-aliasing filter can be of lower order. The decimation, done by a double left bit shift, helps to improve

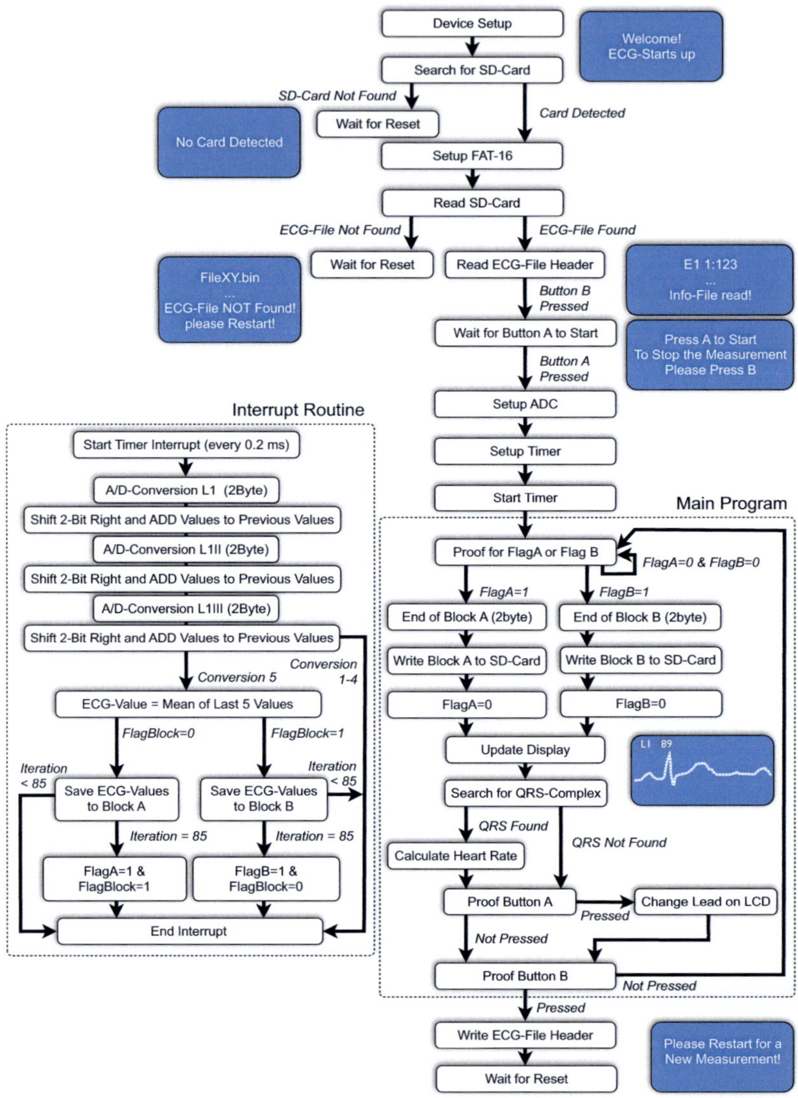

Fig. 4.7. Schematic of the *BSAT-1012* firmware V1.7

the voltage resolution and minimizes the quantification noise. To realise the over-sampling and decimation, in the first four interrupt executions the values are added and bit shifted. In the fifth execution the values are additionally divided by five. The resulting value is saved to one of two block variables.

The program works with two variables to save the signal values. Every variable can save up to 255 2-byte values. When the 255-*th* 2 byte value was written to the block, the program adds a 2-byte "End Of Block" (EOB) sign. The whole data block is written to the SD card. As the time to write the data to the card is longer than the time lag[2] between two timer interrupt executions, this task is done outside the interrupt routine. During the write process of one block variable, the newly converted values are stored in the other block variable. Running like this, the program has a time window of 85 ms to write one block to the SD card.

ECG on Screen

After writing the data to the microSD-card, one of the ECG signals is down sampled by 10 and the resolution is decreased to a range of 64 values. The resulting signal is drawn on the display. If the differences of two successive values exceeds a threshold of 15 pixels (about 0.5V), the controller marks a QRS complex. Once a QRS complex was detected the beat detection is omitted for the next 10 values. The distance between two detected beats is measured and the heart rate is calculated. The result is written on the screen.

End of Measurement

To stop a measurement, the button *B* has to be pressed. The measurement is stopped by changing the timer run flag in the timer control register of the micro-controller. The program writes the number of the last written block to the header of the *ECG.bin* file, changes the display and waits for a reset.

4.2.4.3 Data Storage

The recorded data is stored on a microSD card. The card has to be formatted in a FAT-16 file system and involves a file named *ECG.bin*. In this file, the first two blocks (2 x 512 bytes) are reserved for the header. The header contains information

---

[2] A sampling rate of 5 kHz results in a time lag of 0.2 ms between two successive A/D conversions.

on all signals stored in the file. If the file is empty, the whole header consists of zeros. To find the beginning and the end of a signal, the block number of the first and the last block is set in the header as a numeric 32-Bit value. Figure 4.8 shows a schematic of the files' structure with its header- and data blocks.

The ECG device is not able to generate a new data file, as a "light" version of a FAT-16 driver is implemented in the firmware. However, this is not necessary. The data handling with the analysing software on the computer is more reliable, if the software has to read out one static file and uses the header to distinguish different measurements.

To delete old measurements or to use a new card, the file *ECG.bin* needs to be renewed. A file with 8-bit zero values named *ECG.bin* needs to be generated on the SD-card. The size of the file limits the maximum recording time. For a 12 hour (43200 sec) recording, a file of 259.2 MB is necessary, as every millisecond 3 leads are converted, generating one 16-Bit value each.

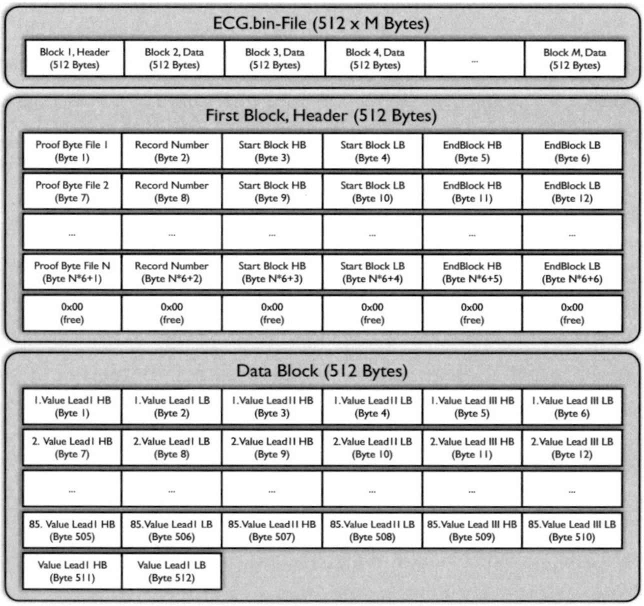

ECG.bin-File (512 x M Bytes)

Block 1, Header (512 Bytes)	Block 2, Data (512 Bytes)	Block 3, Data (512 Bytes)	Block 4, Data (512 Bytes)	...	Block M, Data (512 Bytes)

First Block, Header (512 Bytes)

Proof Byte File 1 (Byte 1)	Record Number (Byte 2)	Start Block HB (Byte 3)	Start Block LB (Byte 4)	EndBlock HB (Byte 5)	EndBlock LB (Byte 6)
Proof Byte File 2 (Byte 7)	Record Number (Byte 8)	Start Block HB (Byte 9)	Start Block LB (Byte 10)	EndBlock HB (Byte 11)	EndBlock LB (Byte 12)
...	...	...	...	...	...
Proof Byte File N (Byte N*6+1)	Record Number (Byte N*6+2)	Start Block HB (Byte N*6+3)	Start Block LB (Byte N*6+4)	EndBlock HB (Byte N*6+5)	EndBlock LB (Byte N*6+6)
0x00 (free)	0x00 (free)	0x00 (free)	0x00 (free)	0x00 (free)	0x00 (free)

Data Block (512 Bytes)

1.Value Lead I HB (Byte 1)	1.Value Lead I LB (Byte 2)	1.Value Lead II HB (Byte 3)	1.Value Lead II LB (Byte 4)	1.Value Lead III HB (Byte 5)	1.Value Lead III LB (Byte 6)
2. Value Lead I HB (Byte 7)	2.Value Lead I LB (Byte 8)	2.Value Lead II HB (Byte 9)	2.Value Lead II LB (Byte 10)	2.Value Lead III HB (Byte 11)	2.Value Lead III LB (Byte 12)
...	...	...	...	...	...
85. Value Lead I HB (Byte 505)	85.Value Lead I LB (Byte 506)	85.Value Lead II HB (Byte 507)	85.Value Lead II LB (Byte 508)	85.Value Lead III HB (Byte 509)	85.Value Lead III LB (Byte 510)
Value Lead I HB (Byte 511)	Value Lead I LB (Byte 512)				

Fig. 4.8. Schematic description of the data file *"ECG.bin"*

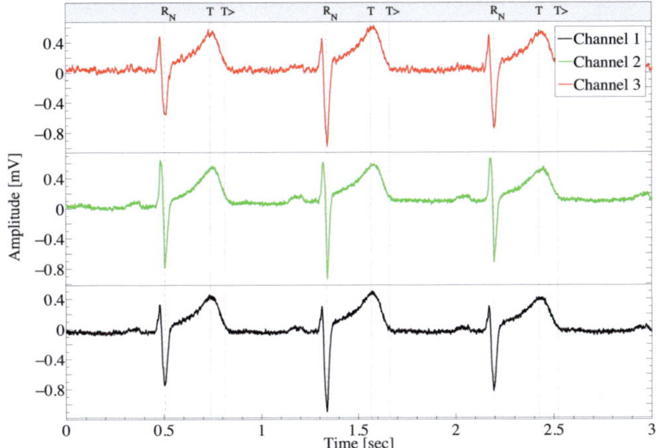

Fig. 4.9. Three lead ECG signal recorded using *BSAT-1012*.

4.3 Test Measurements

Figure 4.9 shows a signal part of an ECG recoding recorded with the *BSAT-1012* device. A clinical diagnostic ECG signal results. All characteristic waves in the ECG, as P wave, QRS complex and T wave are clearly visible.

To get a measure of the quality of the signal acquisition using ECG device *BSAT-1012*, the signal to noise ratio was estimated. SNR is defined as the ratio between the power of a signal and the power of superposed noise. As a recorded ECG signal has unknown superposed noise, the SNR can only be estimated. To do so, the measured signal has to be decomposed into the biosignal and noise. For that purpose, the measured signal was high frequency filtered by a wavelet based filter. Some considerations on wavelet based filtering are made in Section 7.1.2. The mother wavelet for wavelet decomposition was "*Symlet2*", as recommended in [37] for high frequency denoising of ECG signals. Figure 4.10 shows a part of a recorded ECG signal and the corresponding filtered signal.

The difference between the recorded and the denoised signal is assumed to be noise. SNR is calculated using Equation 4.8.

$$SNR = 10 \cdot log_{10} \left(\frac{A_{ECG}}{A_{noise}} \right)^2 \qquad (4.8)$$

Where A represents the Root Mean Square (RMS) of the amplitude of the signal:

$$A_{ECG} = \frac{1}{M} \cdot \sqrt{\sum_{i=1}^{M} ECG_{den}(i)^2} \qquad (4.9)$$

M is the number of sample points of the signal. The result is given in decibel (dB). To test the signal quality, 10 ECGs from 10 different subjects have been analysed regarding the SNR. Signals had a duration of 25 minutes. The results of all 10 measurements are shown in Table 4.1 .

Table 4.1. SNR-values of 10 ECG recordings.

Recording	SNR
1	39.04 dB
2	39.88 dB
3	43.34 dB
4	42.38 dB
5	39,08 dB
6	41,36 dB
7	39,24 dB
8	41,46 dB
9	42,56 dB
10	38,38 dB
Mean	40,67 dB

Mean SNR, $(\overline{SNR})$ of the device is about 40 dB. This is a good result for an ECG device. However, the results are highly dependent on the denoising of the signal. A change in the filter settings leads to a change in SNR. As the ground truth is not available, the denoising is the only way to estimate the SNR. In the noise signal some small peaks can be seen in the region of the QRS complexes. The filtering is strong, which leads to conservative SNR estimation.

As the ECG signal is a biosignal with natural changes over time, the results can be dependent on the signal length. For calculation of the SNR, ECG signals with equal length (25 minutes) have been used. SNR estimation showed similar results in all 10 recordings. A reliable $\overline{SNR}$-value can be assumed.

The visual ECG quality is also high. Diagnostic relevant information are extractable out of the recordings of *BSAT-1012* ECG recorder.

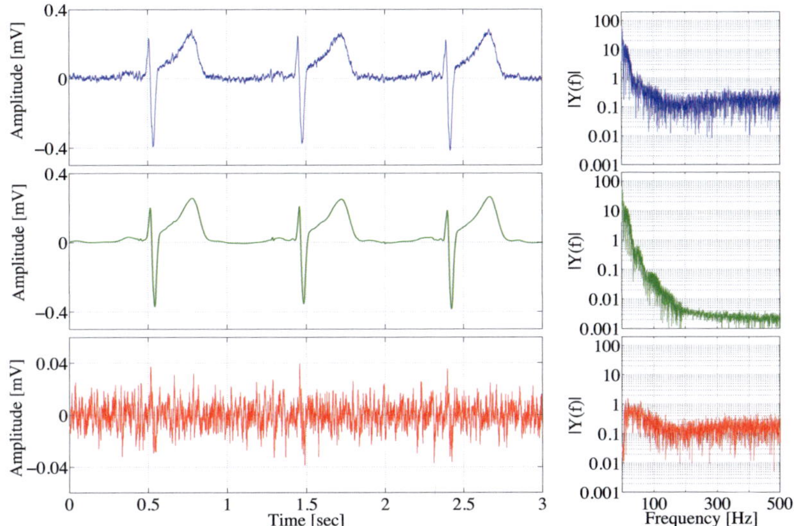

Fig. 4.10. Measured ECG- signal (blue) decomposed by wavelet filter into an ECG signal (green) and a noise signal (red) to estimate the SNR. Corresponding frequency spectrum in the right axes.

Time series on RR- and QT intervals, as well as T wave morphology describing parameters, resulting from exercise tests on a bicycle ergometer, recorded with *BSAT-1012* have been generated and analysed. This investigation is introduced in Chapter 8.4 of this thesis.

5

ECG Databases

5.1 Introduction

In this research work ECG data from several databases have been used. Table 5.1 gives a short overview on the used datasets. The denotation can be seen in the left column of Table 5.1. The data was chosen to satisfy requirements of the research projects they have been used for. They come from different groups and have different basic properties, as for example, sample rate or resolution. Of course, also the underlying subjects are different. Only the THEW- and the IBT-Exercise-study assume healthy subjects, for example. In this section the datasets are introduced. All datasets have been transferred to the Unisens 2.0 format [49] and have been analysed by the IBT software toolbox BSAT introduced in Section 7.6.

5.2 MIT QT-Database

This study is available at the Physionet database [50]. The study named *MIT QT-Database* consists of 105·15 minute, two channel ECG recordings. The tapes have been chosen from Holter ECG data of other studies in the Physionet database, to include a broad variety of QRS and ST-T morphologies. The database was designed to evaluate algorithms that detect ECG waveform boundaries [51].

30 to 100 representative beats were annotated by an expert. As in this research work the annotation of the T wave is in focus, the expert annotation of the T peak and end was analysed.

Signals are recorded with a sample frequency of 250 Hz. In this work, the study is labelled as *MIT QT-Database*.

Table 5.1. Short description of datasets and denotation used in this work

Database	Description
MIT QT-Database	ECG, 2 leads 94 recordings used, reference annotation of QRS complexes and T waves
THEW tQT_I-study	ECG 3 leads, 34 subjects used, 3 times 6 h recordings for each subject, drug safety study (placebo, Moxifloxacin and unknown compound)
THEW tQT_{II}-study	ECG 12 leads, 57 subjects used, 2 times 24 h recordings for each subject, drug safety study (placebo and Moxifloxacin)
Myocarditis-study	ECG, blood pressure and respiration, 1 subject, 7 about 1 h recordings, during therapy, one week intervals
IBT-Exercise-study	ECG 3 leads, 10 healthy subjects, 1 about 45 min. recording for each subject, exercise test (bicycle ergometer)
MIT-BIH Arrhythmia Database	ECG 8 leads, 47 subjects, 48 recordings of 30 min duration each, suffering from Premature Atrial Contraction (PAC) and Premature Ventricular Contraction (PVC).
PVC-study	ECG 2 leads, 56 subjects 1 h recording each subject, suffering from PAC and PVC

5.3 Telemetric and Holter ECG Warehouse

The Telemetric and Holter ECG Warehouse (THEW) is hosted by the centre for Quantitative Electrocardiography and Cardiac Safety at the University of Rochester Medical Centre, Rochester, NY, USA. "The objective of the Centre for Quantitative Electrocardiography and Cardiac Safety and its Telemetric and Holter ECG Warehouse (THEW) is to provide access to electrocardiographic data to research organizations for the design and validation of analytic methods to advance the field of quantitative electrocardiography with a strong focus on cardiac safety."[52]

5.3.1 THEW Thorough QT Study 1

THEW data description adapted from [53]:

Study Design: The study was a double-blind, randomised, 5-way crossover study. The study consisted of a single cohort of 35 subjects. Each subject was randomised to one of 10 possible sequences of treatment and received a single dose of Unknown Compound (UnC): 30 mg / 100 mg and 300 mg. Placebo or Moxifloxacin was given on five occasions, each being 10-14 days apart. During each study period subjects were resident for two nights/three days and a follow-up visit occurred 10-14 days after the final administration. A run-in day, independent from the randomisation was included in period 3 requiring an additional one night/one day

residency during which subjects were administered a single blinded placebo dose.
Inclusion Criteria: Healthy male and female subjects, between age 18 to 55 years, body mass index is approximatively 18 to 30 kg/m^2 and total body weight >50 kg.
Study Treatment: A single oral dose of UnC tablets (30, 100 and 300 mg), placebo or a single Moxifloxacin tablet (400 mg) was administered on five separate occasions with 10-14 days between each study period.
The internal identification number at THEW is: E-HOL-03-102-005.

In this thesis 34 participants were investigated. For each of the participants three 10 hour ECG recordings dosed with placebo, Moxifloxacin and an unknown compound have been analysed. For evaluation a time interval from dose to four hours post dose was in focus.
Signals are recorded with a sample frequency of 200 Hz. In this work, the study is labelled as *THEW tQT$_1$-study*.

5.3.2 THEW Thorough QT Study 2

THEW data description adapted from [54]:
Study Design: This is a single-centre, single-dose, randomized, double-blind, double-dummy, placebo-controlled, active-comparator, four-way crossover study. Each subject received a single dose of each of treatment (e.g., A, B, C, and D) by random assignment to one of four sequences (ABCD, BDAC, CADB, or DCBA). These treatments will be separated by a washout period of at least 7 days.
The four treatments consist of the following:

1. Treatment A: - Therapeutic dose of RO X treatment (**data not provided**)
2. Treatment B: - Supratherapeutic dose of RO X treatment (**data not provided**)
3. Treatment C: - Active comparator treatment - 400 mg Moxifloxacin capsule
4. Treatment D: - Placebo treatment

The total duration of the study is up to 9 weeks (from screening through study completion) for each enrolled subject as follows:

- Screening: Up to 4 weeks
- Dosing periods: 4 weeks (1 week washout period between each dosing)
- Follow-up: 7 (+3) days after last dosing (study completion)

The end of the trial is defined as the date of the last visit or receipt of last data point for statistical analysis of the last subject undergoing the trial.

The internal identification number at THEW is: E-HOL-12-0140-008.

In this thesis the THEW thorough QT Study # 2 was used with treatment A (Moxifloxacin) and the placebo treatment D. The recordings have a duration of 24 hour each and the measurement of the blood plasma concentration of the treatment Moxifloxacin was used for evaluation. In total recordings of 57 participants of the study were investigated in this research work.

Signals are recorded with a sample frequency of 1000 Hz. The study is labelled as *THEW tQT_{II}-study* in this work.

5.4 ECG-Study of Myocarditis Patients

This study was recorded to investigate patients suffering from myocarditis. Measurement of single lead ECG, continuous blood pressure and respiration were made. The recordings have a duration of 30 to 45 minutes. All signals are sampled with a frequency of 1000 Hz [55].

In the available part of the study, one patient has been recorded during the course of his disease. Recordings have been made with a time lag of about 7 days. In total 8 recordings starting from the day of committal to hospital are available. Further personal or clinical information are not available for the patient.

The study has been provided by courtesy of Prof. Dr.-Ing. habil. Hagen Malberg of the Institut for Biomedical Engineering at technical University of Dresden. The Study is labelled as Myocarditis-study.

5.5 IBT-Exercise-Study

To investigate the influence of the heart rate to the repolarisation process of the human heart, stress test ECG data has been recorded. The study involves 10 healthy, male subjects with a mean age of 24.8 years (SD:$\pm$2.49 yr.). The subjects made exercises on a bicycle ergometer with resting periods for about 45 minutes each. The recordings have been made at the Institute of Biomedical Engineering (IBT) of the Karlsruhe Institute of Technology (KIT), using ECG device *BSAT-1012*. Exercise was started few minutes after starting the measurement. First, heart rate was tried to get as high as possible. Participants decided when a maximum was reached. A resting sequence followed until heart rate was back on baseline or did not significantly change for some minutes. Next, an exercise was started again and the heart rate was increased by half of the range of the previous exercise. Once this maximum was reached, again a resting sequence followed. In cycle three and for same was done, for a heart rate change of 0.75 times of the first heart rate range and the full range again. Figure 5.1 elucidates the time course of the heart rate.

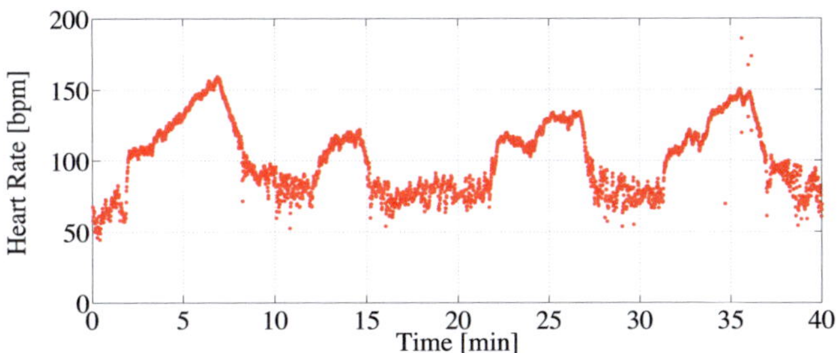

Fig. 5.1. Time course of heart rate in IBT-Exercise-Study for participant P_2

In this work, the study is labelled as *IBT-Exercise-Study*. Recordings of the participants are labelled as P_1 to P_{10}.

5.6 MIT-BIH Arrhythmia Database

Description adopted from the Physionet Website [56, 57]:
The MIT-BIH arrhythmia database contains 48 half-hour excerpts of ambulatory ECG recordings, obtained from 47 subjects studied by the BIH Arrhythmia Laboratory. 23 recordings were chosen at random from a set of 4000 24-hour ambulatory ECG recordings collected from a mixed population. The remaining 25 recordings were selected from the same set to include less common but clinically significant arrhythmias that would not be well-represented in a small random sample.
Technical data:

- Sample frequency 360 Hz
- Resolution 11-bit over 10 mV range
- Two channels each recording

- Annotation were made by two or more cardiologists

In this work the study is label as *MIT-BIH arrhythmia database*.

5.7 Database of Premature Ventricular Contraction

In this study, 56 one hour ECG recordings are available. The ECGs involve a significant number of premature ventricular contraction (PVC). The history of the clinical diagnostics of the patients recorded in this study is not available. The records are sampled with a sample frequency of 500 Hz and involve eight ECG leads, each.

The study has been provided by courtesy of Biosigna Munich. In this work the study is label as *PVC-study*.

6

State of the Art in ECG Signal Processing

Since the physician William Einthoven developed the first ECG for clinical practice in 1903, the recording of the heart signals has become one of the most important diagnostic procedures in clinical practice. A lot of different approaches have been introduced to analyse ECG signals since the beginning of the clinical ECG recording in 1903. However, the rapid development of computers beginning in the 1980s led to new technologies to record and moreover analyse ECG signals using the now available computing power. In this section, a short overview over the computer assessed signal processing of ECG signals is given. The introduction is based on common methods used in recent years. The author gives only a short overview in this thesis. References can be used to find more detailed information on the methods. Clifford et al. in their book 'Advanced Methods and Tools for ECG Data Analysis' [58] present a wide and detailed compendium on ECG signal processing. A compendium on bioelectrical signal processing in general is presented by L. Sörnmo and P. Laguna in 'Bioelectrical Signal Processing in Cardiac and Neurological Applications' [59]. Both books are recommended for further study in ECG signal processing by the author.

In this section, the focus of ECG signal processing is on ECG delineation and beat classification.

6.1 QRS Detection

The detection of the R peaks in the surface ECG is the most important delineation task, since it is a precondition for most ECG analyses. A good overview of the principles of QRS detection is given by B.U. Köhler in [60]. Some of the assumptions made there have been summarized in this section.

In 1985, Pan and Tompkins introduced one of the most often used online R peak detection method [61]. This real time ECG delineation algorithm detects a QRS complex by the digital analysis of slope, amplitude and width of an ECG signal segment. False detection due to interference signals are minimized by a band pass filter. All used thresholds automatically adapt to the ECG signal to provide a reliable method independent from signal quality and electrode position.

P. Laguna et al. present an ECG delineation algorithm working on several ECG leads in parallel in [62]. The delineation is separately carried out in each lead similar to the method obtained in [61]. The detected QRS complexes are then compared and a decision whether a beat was found or not is made considering information of all leads. The algorithm works more robust in noisy ECG signals.

In the last two decades, the Wavelet Transform (WT) became a commonly used method for QRS detection in surface ECG signals. Feature extraction, in terms of irregular structures in the signal, using WT and estimation of local Lipschitz exponents was first introduced by S. Mallet and W.L. Hwang in [63]. Based on this work, approaches have been developed to detect the R peaks in the ECG, by scanning relevant scales of the WT. In [64] Li et al. detect the R peaks using Lipschitz regulatory and heuristic decision rules. Small adaptations have been made to simplify the approach shown in [64], so that the analysis can be carried out in real time[65, 66].

The WT of the signal provides a well working method to detect the high frequency peaks of the QRS complex. The closely related Wavelet filter banks are one of the most common approaches of QRS delineation in the surface ECG today. Khawaja presented a method using Wavelet filter banks in combination with the *Haar*-mother wavelet to detect the QRS regions in the surface ECG [37, 67]. A similar work is presented by Martinez in [68]. However, differences exist in the analysis of the detail coefficients, the used decomposition levels and the thresholds.

The combination of frequency and time related information makes the Wavelet approach to a reliable delineation tool. The results of delineation (sensitivity and positive prediction value) of these algorithms in detecting QRS complexes lies above 99% in most publications. A detailed description of a Wavelet based R peak detection is presented in Section 7.2.1.

In recent years, approaches based on neural networks have been introduced. Those approaches have to be trained. Some can classify the detected beats. These and some more investigations are described in detail in [58, 59, 60].

6.2 T-Wave Delineation

The delineation of the T wave is a challenging problem. The end of the T wave is a sample near the baseline and therefore the signal to noise ratio is very low. However, the T wave boundaries are extremely important in the clinical routine. Thus, a reliable automatic delineation of the T wave has been in focus for more than 6 decades now. The first approach in finding the end of the T wave is based on a tangential method. In 1952, Lepeschkin et al. have introduced a method that defines a tangent to the maximum slope of the falling edge (positive T wave) of the T wave [69]. The crossing of this tangent with the baseline is defined as T_{end}. This approach works well for signals with only moderate baseline wander. Thus, the method was adapted by the author to delineate the subject specific T waves introduced in Section 7.3.1.

Another approach introduced by Zhang et al. is based on the analysis of an indicator signal calculated from the area covered by the T wave [70]. The indicator signal is the result of the area under the T wave in a sliding window. The area under the curve is not limited by the baseline, but by the lowest sample inside the window. An extremum in the resulting indicator curve has chosen to be T_{end}. Biphasic T waves are not considered in this approach.

Laguna et al. introduced an ECG delineation algorithm with QRS and T wave delineation in [62], as already mentioned above. The delineation can be performed on multi lead ECG signals. To delineate the T wave, once the differential and once a band pass filtered signal is computed from the ECG signal. Zero crossings of the differential signal mark the T wave peak. Based on this, the boundaries of the T wave can be found in the band pass signal using an adaptive threshold. The results from each lead are taken into account to find the wave boundaries. I.e. the earliest onset, which is not assumed to be an outlier, and the latest offset, which is also not assumed to be an outlier, are taken as the wave boundaries.

Beside the delineation of the wave boundaries based on filtered ECG signals, Wavelet based approaches are very common. In [68], Martínez detects the T wave between two successive R peaks by analysing the wavelet scales 2^4 or 2^5. The wave boundaries and peaks are again found by the zero crossings and extrema in the WT of the corresponding scales. A similar approach is presented by Li et al. in [64] and Khawaja in [37].

6.3 Beat Classification

The delineation of the QRS complexes in an ECG recording provides information on the heart rhythm. To evaluate the heart rhythm clinically, additional information on the type of beat is necessary. In this thesis, normal beats and ectopic beats were in focus. Based on their origin in the ventricles or in the atrium, ectopic beats can be separated into Premature Ventricular Contraction (PVC) and Premature Atrial Contraction (PAC). This section introduces some published methods to detect and distinguish ectopic beats.

To detect ectopic beats in a surface ECG, rhythmical features of the recording have to be analysed. The rhythmical characteristic is significantly changed by both, PAC and PVC. The RR interval before the ectopic beat (so-called coupling interval) is shorter than the normal rhythms of previous beats. Moreover, the first subsequent RR interval after the PAC (so-called compensatory pause) is shorter than the following RR intervals. This fact makes a decision based on rhythmical features possible. The morphology of the QRS complex of the PVCs is altered, whereas the morphology of PACs is equal compared to the morphology of normal heart beats. The number of publications discussing the detection of PACs is smaller compared to the publications presenting methods to classify PVCs. However, there are also publications discussing both types in common. Krasteva et al. presents a method based on inter beat differences of the RR intervals of 5 successive heart beats in [71]. To distinguish PAC from PVC, additional morphological descriptors are introduced. These are based on the width of the QRS complex, area under the curve and an angle in the vectorcardiogramm. To classify the beats, subsequent decisions are made based on the descriptors and predefined thresholds. The introduced algorithms do not have to be trained.

A comparison of two approaches for classifying ventricular beats is made in [72] by Millet et al. . The presented method can distinguish normal and ventricular beats only. The first approach defines parameters only in the signal sequence between the onset of the QRS complex and the offset of the QRS complex. The second approach defines a 150 ms window, in which a QRS complex lies, and calculates the parameters from the signal inside the window. The parameters are mainly based on area, QRS width and extrema of the signal sequence. A logistic regression is applied to a statistical study which results from the training database. Two linear discriminant functions result, one for each approach. The linear combination of all features leads to the class of the corresponding beat. Millet et al. came to the conclusion that a previous calculation of QRS on- and offset is not necessary for beat classification.

Maier et al. analysed two different strategies for non-supervised classification of QRS complexes in [73]. The first approach uses a hierarchical cluster analysis. This is based on three different feature sets of Fourier coefficients, discrete cosine transform and coefficients of an Hermite series expansion. The beats are combined in clusters, based on the distance in the feature space. Finally, the majority of a cluster decides to which class the beats belong. The second approach is based on cross-correlation of the beats. In a first step, single beats are defined as templates and are cross-correlated to the other beats. In a second correlation step, an average template of the classes found in the first step is generated and again cross-correlated to the beats. After the second correlation step, the beats are combined into clusters. Again, the majority of a cluster decides to which class the beats belong. It has been concluded that the correlation based approach is superior to the hierarchical cluster analysis. The set of FFT based coefficients seemed to be the best choice in the first approach.

The separation into normal beats, PACs and PVCs is mostly based on both rhythmical features and additionally morphology based features. While the rhythmical features are very similar in most approaches, morphological features differ clearly among the publications. The classification process itself is then done by very different approaches as seen in the publications above. Classification methods using supervised and non-supervised learning strategies are very common in ECG beat classification today. In [58], a detailed introduction on supervised and non supervised learning for ECG classification is given. A further compendium on different classification strategies can be found in *'Advances in Cardiac Signal Processing'* [74].

Part II

Methods

ECG Signal Processing

7.1 ECG Denoising

The recording of ECG signals on the humans body surface is corrupted by noise
and artefacts. These have to be suppressed during the analogue data acquisition
or afterwards in a digital signal preprocessing. This chapter shows methods which
were used to denoise ECG signals in the research projects of this work.

Depending on the task and thus the frequency characteristics, three digital filters
with different cut-off frequency (f_N) are available:

- High pass filter to reduce baseline wander $f_N < 0.3Hz$
- Notch filter to suppress power line influence $f_N = 50Hz$ (*European Union*)
- Low pass filter to suppress high frequency noise in the signal $f_N > 80 - 120Hz$

Depending on the quality of the ECG recording and the desired signal process-
ing, the cut-off frequencies of the filters can be set individually. The developed
BioSignal Analysis Toolbox BSAT described in Section 7.6 has a Graphical User
Interface (GUI) for an individual filter design. Based on the mathematical back-
ground, two classes of filters are available in the GUI. The first class uses the
Fast Fourier Transform (FFT), while the second class uses the Stationary Wavelet
Transformation (SWT). In both classes high, low and band pass filters are avail-
able.

7.1.1 Fourier Based Filter

Filter based on the Discrete Fourier Transformation (DFT) are commonly used
for off-line biosignal analysis, because of an easy implementation and the large
computing power available today. In signal processing, filters often have to be real-
time capable. In this case, a windowing is necessary. In case of an off-line analysis

the method remains more simple, as a windowing is not necessary. In case of very long ECG signals the windowing becomes important again. However, signals are transformed by the DFT into the frequency domain. The utilized algorithm is the FFT as described in Section 3.1.2.

To filter a one dimensional signal vector $\boldsymbol{X}$ with a length of N elements, a DFT represented in Equation 7.1 is performed.

$$\underline{\boldsymbol{Y}}(k) = \sum_{n=0}^{N-1} \boldsymbol{X}(n)e^{-j\frac{2\pi kn}{N}} \qquad k = 0, ..., N-1 \tag{7.1}$$

This transformation leads to a complex vector $\underline{\boldsymbol{Y}}$ of length N. The items in this vector represent the frequencies of the signal $\boldsymbol{X}$. To design a filter, all coefficients representing frequencies which need to be damped are decreased or set to zero. As the spectrum in vector $\underline{\boldsymbol{Y}}$ is mirrored, each two boundaries have to be found to design a low or high pass filter. Figure 7.1 illustrates the frequency bands for the three types of filters. The attenuating band of a high pass filter is represented by the zero values of the blue line, the ones of a notch filter by the zero values of the red line and the ones of a low pass by the zero values of the green line. To damp frequencies in the attenuating band of the filters, boundary frequencies in $\underline{\boldsymbol{Y}}$ have to be calculated. Equation 7.2 shows the calculation for a value representing a frequency in the left half of the spectrum, while the corresponding value in the right half of $\underline{\boldsymbol{Y}}$ is given by 7.3.

$$b_L = \left\lceil \frac{f_x \cdot N}{f_s} \right\rceil + 1 \tag{7.2}$$

$$b_R = N - \left\lceil \frac{f_x \cdot N}{f_s} \right\rceil + 2 \tag{7.3}$$

With the knowledge of the indices of the cut-off frequency of the filter, the corresponding values in the attenuating band can be set to zero. This can be realized by a multiplication vector $\boldsymbol{W}$. $\tilde{\underline{\boldsymbol{Y}}}$ is then the new complex spectrum:

$$\tilde{\underline{\boldsymbol{Y}}} = \boldsymbol{W} \cdot \underline{\boldsymbol{Y}} \tag{7.4}$$

In the last step of the filtering, the modified complex coefficients of the DFT $\tilde{\underline{\boldsymbol{Y}}}$ are transformed by an inverse Discrete Fourier Transform (iDFT) (Equation 7.5) to the signal $\tilde{X}$ in the time domain. For the computation, the algorithm of the inverse Fast Fourier Transform (iFFT) is used.

$$\tilde{X}(n) = \sum_{n=0}^{N-1} \tilde{Y}(k)e^{j\frac{2\pi kn}{N}} \tag{7.5}$$

Figure 7.2 shows an ECG signal and its Fourier spectrum before and after denoising. FFT filters of high pass, low pass and a notch filter have been applied using BSAT.

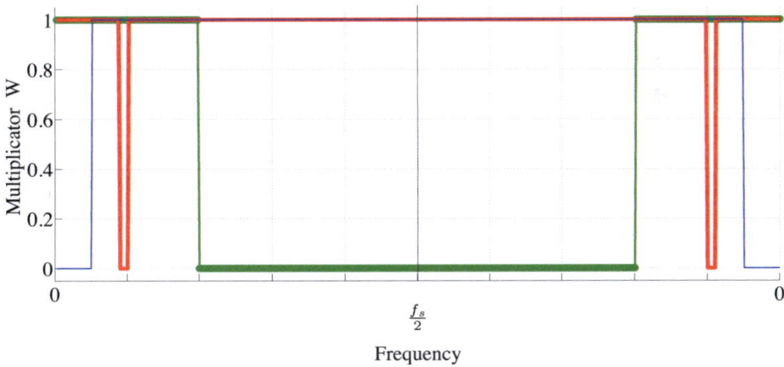

Fig. 7.1. Examples of multiplier vectors for high, low and band pass FFT filters. The zero values of the coloured lines illustrate values of $\tilde{Y}$ belonging to the attenuating bands of the filters. Blue: high pass filter, red: notch filter and green: low pass filter. As the figure is for illustration purpose, the boundaries of the filters do not represent an ideal ECG filter.

7.1.2 Wavelet Based Filter

The basics of wavelet based ECG filtering has been adapted from the work of Khawaja [37]. The algorithm has been implemented in BSAT to be used during signal pre-processing of ECG data. Only a short overview is given in this section, further information can be found in [37] and [33].

The Discrete Wavelet Transform (DWT), introduced in Section 3.2, decomposes the signal X into a set of signals representing a multi-scale filter bank. As mentioned in Section 3.2.2, approximation coefficients of level l can be split into detail and approximation coefficients fo level $l + 1$. This results into a low pass filtered signal for the approximation coefficients and a band pass filtered signal for the

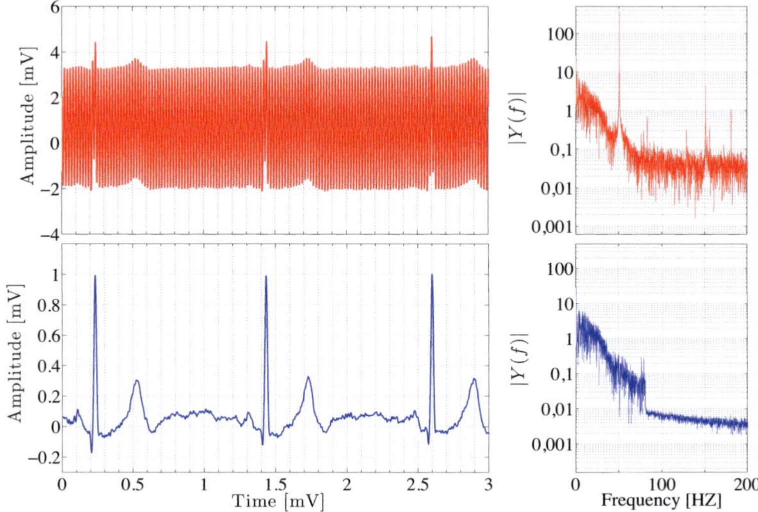

Fig. 7.2. ECG signal with strong power line noise (red) and the same signal after denoising with a 0.3 Hz high pass, a 50 Hz notch filter and a low pass filter of 80 Hz. The corresponding frequency spectrum can be seen on the right. The original spectrum (red) has a high peak at the 50 Hz position. In the filtered spectrum the peak is missing (blue line). The spectrum above 80 Hz is zero in the filtered case (blue). The effects of the high pass cannot be seen cause of the axes scaling.

detail coefficients. A modification of the detail and approximation coefficients and the reconstruction afterwards, change the signal in the corresponding frequency range. This is the basic idea of the wavelet filter approach. In this work the Stationary Wavelet Transform (SWT) is used, which preserves the length of the decomposed signal in all detail and approximation coefficient signals.

Using this method, specific high and low pass filters can be created. Band pass filters for ECG denoising using the DWT are not common. Normally power-line influences are cancelled by notch filters in ECG signal processing. The frequency bands of a wavelet filter bank are dependent on the sampling frequency. A small band in the region of the power-line is normally not available. Detail coefficients of level 3 e.g. of a signal sampled at 1000 Hz, represent a frequency band between 31,25 Hz to 62.5 Hz. 50 Hz power-line noise lies inside this band. Nevertheless, important signal information of the ECG is also present in this band.

Baseline Wander

The recorded ECG signal suffers normally from a variance in the baseline. This variance is often generated during recording by unsymmetrical changes of the impedance of the electrode-skin contact. However, this movement of the baseline comes not from a physiological effect of the heart. Hence, it has to be filtered out.

An example is given to elucidate the method of wavelet based high pass filtering by the cancellation of baseline wander. To cancel low frequency noise in an ECG signal X, sampled at 1000 Hz, a single-level one-dimensional wavelet decomposition up to level 11 is performed as described in Section 3.2. This results in detail coefficients D_{11} and approximate coefficients A_{11}. D_{11} involves information of the frequency band 0.24 Hz to 0.49 Hz. A_{11} involves information below 0.24 Hz. To get rid of the baseline wander, which is assumed to be below 0.24 Hz, the approximation coefficients A_{11} are set to zero and the signal is reconstructed. The reconstruction of the signal is done by a iSWT.

High Frequency Noise

To cancel high frequency noise, the decomposition of the ECG signal is done by the SWT to receive detail- and approximation coefficients. Frequency bands assumed not to contain any biosignal can be deleted by setting all detail coefficients to zero. In frequency bands involving both, signal- and noise information, all detail coefficients D_l below a threshold Tr are set to zero. The threshold depends on the filter design. Two different approaches for thresholding are common:

- **Hard Thresholding**: All values d_l of detail coefficient vector D_l below Tr are set to zero.

$$\tilde{d}_l = \begin{cases} d_l, & \text{case} |d_l| \geq Tr \\ 0, & \text{case} |d_l| < Tr \end{cases} \tag{7.6}$$

- **Soft Thresholding**: All values d_l of detail coefficient vector D_l below Tr are set to zero, while values above or equal Tr are set to the distance between d_l and T_r.

$$\tilde{d_l} = \begin{cases} sign(d_l)(|d_l| - Tr), & \text{case}\,|d_l| \geq Tr \\ 0, & \text{case}\,|d_l| < Tr \end{cases} \tag{7.7}$$

Mother Wavelet

In BSAT, the type of mother wavelet can be chosen. Most of the mother wavelets available in Matlab have been implemented. In [37] several mother wavelets have been tested. The author of [37] recommends the *Daubechies11* mother wavelet for baseline wander cancellation and the *Symlet2* mother wavelet for high frequency noise.

7.2 Fiducial Point Detection

The decomposition of an ECG signal by the WT provides a reliable possibility to delineate the QRS complex of the ECG. Today's computational power allows the real time calculation of this transformation even in on-line measurement systems which makes it one of the most common methods [75].

To delineate QRS complexes and T waves in the ECG, the algorithm introduced here detects the R peaks by analysing the detail coefficients of a SWT decomposition. The R peak detection using the SWT bases upon the ideas presented in [37]. Once the R peaks have been detected, QRS complexes can be marked by finding corresponding Q- and S peaks. A classification of the QRS complexes by a Support Vector Machine (SVM) is used to distinguish between Premature Ventricular Contraction (PVC), Premature Atrial Contraction (PAC) and normal heart beats. The classification is described in Section 7.4. The T wave delineation is done by a correlation based method described in Section 7.3. Figure 7.3 shows a schematic diagram of the fiducial point detection process.

7.2.1 QRS Delineation

Pre-Filtering

The major frequencies of QRS complexes in an ECG lie in a frequency band between 2 Hz and 40 Hz. Hence it is beneficial to first filter the recorded ECG. A high pass FFT filter of 2 Hz is used to suppress baseline wander in the signal. This might change the shape of the P and T wave, but a significant modification of the characteristic shape of the QRS complex is not observed. A low pass FFT filter

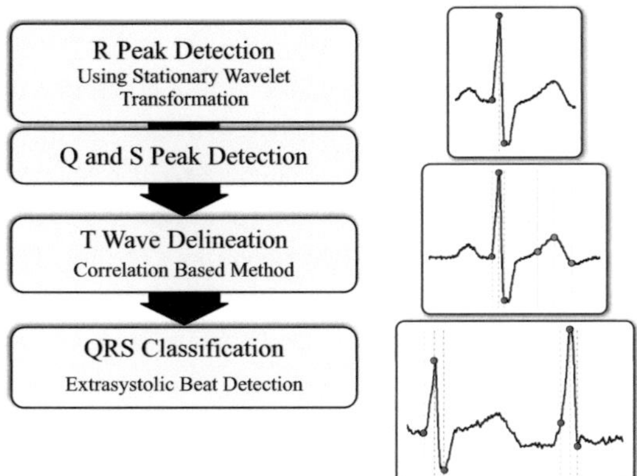

Fig. 7.3. Schematic of Fiducial Point Detection

with a cut-off frequency of 45 Hz is used to suppress high frequency noise in the signal. This might slightly round the peaks of the QRS complexes, but won't influence the steep edges of the QRS complexes. Steep edges of the QRS complexes are mainly responsible for the delineation process. Any undesired change in the waveform by using these filters for delineation can be turned back after delineation is finished.

Wavelet Transformation

The original recorded ECG signal is decomposed by a SWT into several signals of detail and approximation coefficients. In this chapter as mother wavelet the *Haar*-wavelet is used. The detail coefficients $\boldsymbol{D}$, used for R peak detection are dependent on the sampling frequency f_s of the recorded signal. Detail coefficients of every level l involve frequencies of a frequency band between f_{N+} and f_{N-}. For an ECG signal recorded at $f_s = 1000\,\text{Hz}$, $\boldsymbol{D}_6$ represents the detail coefficients of the frequency band involving $f_q = 15\,\text{Hz}$. The boundaries of the frequency band can be calculated in dependency of the sampling frequency f_s:

$$f_{N+}(l) = \frac{f_s}{2} \cdot 2^{-l} \tag{7.8}$$

$$f_{N-}(l) = \frac{f_s}{2} \cdot 2^{-l+1} \tag{7.9}$$

For QRS delineation, a decomposition by the SWT has to be done up to the level l_q covering f_q. The resulting signal of detail coefficients D_{l_q} is analysed to find regions potentially involving a QRS complex. It turned out that the reliability of the algorithm can in some cases be improved by analysing the differential signal of D_{l_q} (dD_{l_q}). dD_{l_q} is computed and analysed equal to D_{l_q}. Computation time is quite short for this step compared to the SWT decomposition and thus region analysis of both signals can be done without a significant loss of time. Decision whether the results based on D_{lq} or dD_{l_q} are used is made by analysing the returned RR time series in a later step. The following explanations are given for D_{l_q} only, but are similar for dD_{l_q}.

QRS Region Detection

At the position of a QRS complex, a sharp peak is found in D_{l_q}. The *Haar*-wavelet has a high correlation with the steep increasing and decreasing edges of the QRS complexes. To find all potential regions of a QRS complex, areas in the signal having values greater than a threshold need to be identified. Artefacts, noise and strong baseline wander, as well as the amplification factor of the ECG device have an influence on the signal. Thus, a threshold has to be adaptable to the current shape of the ECG signal to ensure the finding of all potential QRS regions.

To adapt the threshold to the current signal shape, a Root Mean Square (*RMS*) of the ECG signal in a sliding window of two seconds is performed (Equation 7.10). In case of only few QRS complexes inside the siding window, the *RMS*-value ς_i is highly influenced by a change of the number of QRS complexes inside the window. As the number is heart rate dependent and cannot be calculated at this state, a heart rate dependent window size is not possible. However, the window size should be chosen to involve at least one QRS complex surely. An increased window size leads to a decreased signal adaptivity. This can be a problem in case of motion artefacts, e.g. as the adaptation of the threshold might be too slow. Keeping all this in mind, it turned out that a window size of two seconds is a good compromise for ECG signals of healthy subjects. Using this window size, at least one QRS complex is inside the window. For heart rates below 60 bpm even two QRS complexes are surely inside. Furthermore, changes due to artefacts should increase the *RMS* within less than one second.

$$\varsigma(i) = \sqrt{\frac{1}{2 \cdot ws + 1} \cdot \sum_{n=i-ws}^{i+ws} \boldsymbol{D}_N(i+n)^2} \qquad (7.10)$$

In this equation, i represents the sample number of the current position in the ECG signal and $2 \cdot ws + 1$ is equal to the number of samples inside the sliding window. Calculation of ς in a high resolution Holter ECG signal leads to a significant increase in computation time. For this reason, the algorithm does not calculate a real *RMS*. The difference between the *RMS*-signal and a mean value filter of the absolute values of the detail coefficients, as presented in Equation 10.1, are negligible. As the threshold has to be adaptive, a multiplication factor M is introduced.

$$\boldsymbol{\tau}_M(i) = \frac{M}{2 \cdot ws + 1} \cdot \sum_{n=i-ws}^{i+ws} |\boldsymbol{D}_N(i+n)| \qquad (7.11)$$

Figure 7.4 shows a 500 Hz sampled ECG signal and the belonging detail coefficients D_5. Further more ς, $\boldsymbol{\tau}_{1.75}$ and $\boldsymbol{\tau}_{7.5}$ are drawn. The difference of the $\boldsymbol{\tau}_{1.75}$ to the ς is not crucial for the region detection.

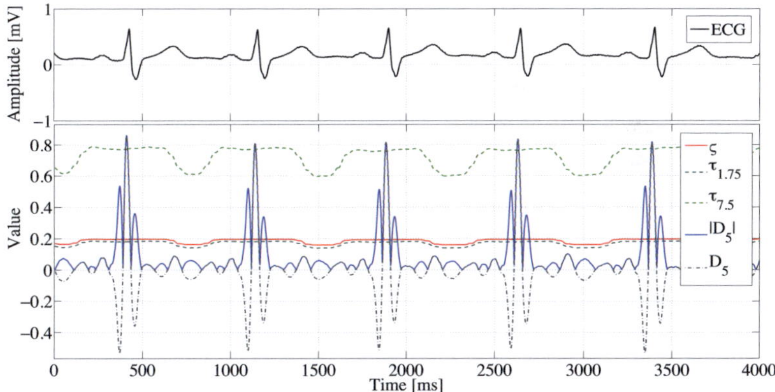

Fig. 7.4. ECG signal (black) and the corresponding detail coefficients of the level five, representing the frequency band involving the frequency of $15 Hz$ (dashed grey). The blue signal represents the absolute value of the detail coefficients. Thresholds for QRS region detection can be seen in dashed green. As the threshold is optimized during detection, the two dashed green threshold lines represent the highest and lowest possible thresholds. Red line shows the threshold line calculated by ς. The differences between the red and dashed green signal are marginal.

Threshold Optimization

All parts of the ECG signal with values in $\boldsymbol{D}_{l_q}$ larger than $\boldsymbol{\tau}_M$ are identified as a potential QRS region. Regions having a time distance smaller than 5 ms against each other are combined. The time difference between the first values of two successively detected regions is interpreted as a pseudo RR interval. The detected pseudo RR intervals are combined in vector $\widehat{\boldsymbol{RR}}$.

To improve the quality of the QRS region detection, the optimal multiplication factor M for the threshold has to be found. On the one hand, regions belonging to the P- or T wave are detected by mistake if the threshold is chosen too small. On the other hand, QRS regions might be missed during region detection if the threshold is chosen too high. Altogether, falsely detected or missed regions lead to leaps in the pseudo RR interval vector. As the RR interval is not constant, analysing the variance of $\widehat{\boldsymbol{RR}}$ is not feasible. Thus $\widehat{\boldsymbol{RR}}$ is low pass filtered by a mean value filter with a window length of five pseudo RR intervals. The difference between $\widehat{\boldsymbol{RR}}$ and the mean filtered $\widehat{\boldsymbol{RR}}$ has high values at any leaps in $\widehat{\boldsymbol{RR}}$. Generating the sum of the differences, leads to a value OV representing the amount of leaps in $\widehat{\boldsymbol{RR}}$. OV can be minimized to find the best multiplication factor for the threshold. Equation 7.12 represents the calculation of OV_I.

$$OV_I = \sum_{i=1}^{N} \left| \widehat{RR}_I(i) - \frac{1}{5} \sum_{n=i-2}^{i+2} \widehat{RR}_I(n) \right| \tag{7.12}$$

Figure 7.5 shows all pseudo RR interval time series $\widehat{\boldsymbol{RR}}_I$ of an optimisation process. $\widehat{\boldsymbol{RR}}_4$ of iteration 4 represents the best RR interval vector. The spikes in the curves come from ectopic beats. These remain detectable which is very important. Elimination of potential QRS complexes based on too short RR intervals often leads to a cancellation of ectopic beats. This was the reason for developing a more complex method based on optimizing the thresholds. Figure 7.6 shows a segment of a $\widehat{\boldsymbol{RR}}$ signal, the mean filtered signal and the differences. The peaks in this figure come from an ectopic beat.

OV-values are computed for both, $\boldsymbol{D}_{l_q}$ and $\boldsymbol{dD}_{l_q}$ in all iterations. The decision whether the regions of $\boldsymbol{D}_{l_q}$ or $\boldsymbol{dD}_{l_q}$ are used is also made on the lowest OV-value. Hence, the lowest OV-value indicates the best combination of multiplier and analysis signal.

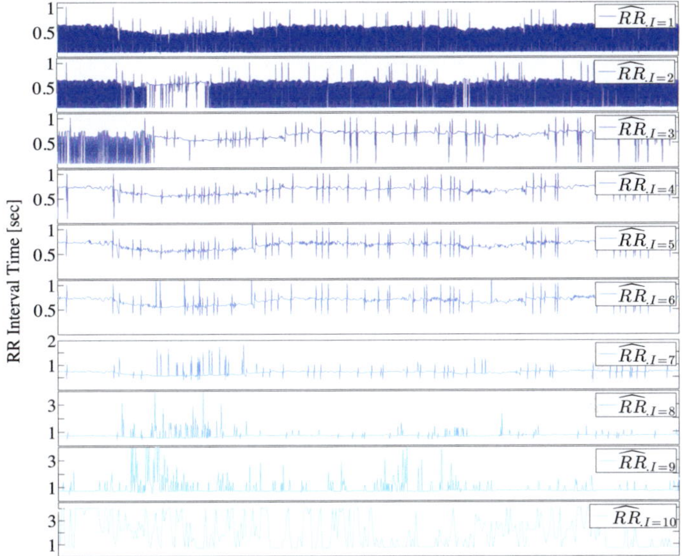

Fig. 7.5. Calculated pseudo RR interval time series coming out of changing the multiplication factor of the threshold signal. It can be seen that for small multiplication factors (lines 1-3) the variance of the signals is high. It decreases significantly by an increasing multiplication factor (line(4-6). Higher values of the multiplication factor (line 7-10) lead to an increasing variance, as missing QRS complexes cause high values in the pseudo RR intervals. The jitters in line 4-6 come from ectopic beats which successfully remain in the QRS detection.

QRS Peak Detection

In every potential QRS region, a search for the R peak is made by analysing the ECG signal. As in this step the correct position of R peak plays a major role, the original ECG signal is FFT filtered by a low-pass of 2 Hz and a high pass of 70 Hz. Detection of the exact position of R peak is done by classic curve sketching. Zero-crossings of the differential ECG signal represent the local maxima and minima. The maximum or minimum having the highest absolute value is assumed to be an R peak. Biphasic QRS complexes might lead to an alteration between the positive and negative peak of the QRS complex during R peak detection. To prevent this, both positions are stored and the description whether the minimum or maximum is used, is decided at the end for all heartbeats in total. The decision is made on the number of occurrence.

Q peaks are found as an extrema of the signal as the last zero-crossing of the dif-

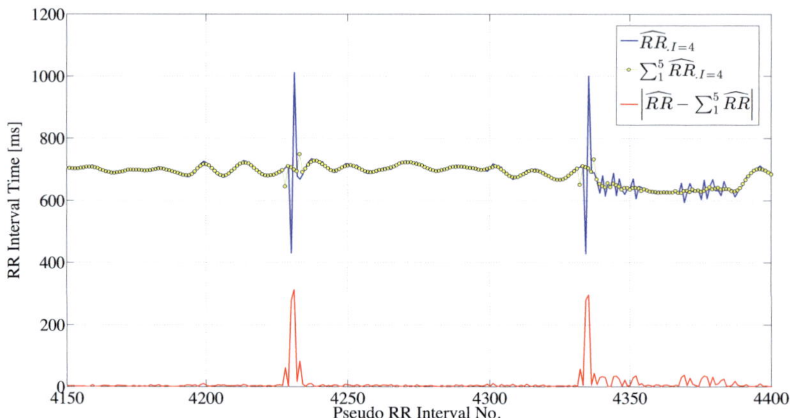

Fig. 7.6. Example of analysing the variance of the pseudo RR intervals. Pseudo RR interval time series of iteration 4 (blue). Smoothed pseudo RR intervals (yellow dotted line). Difference between smoothed and none smoothed signal (red). Peaks in the pseudo RR interval time series lead to a high peak in the difference signal. A measure of the variance for a pseudo RR interval time series is calculated as the sum of all values of the red line.

ferential signal, before an R peak. The S peak is accordingly found by the first extrema after an R peak. Figure 7.7 shows an example of the detection of the peaks of a QRS complex.

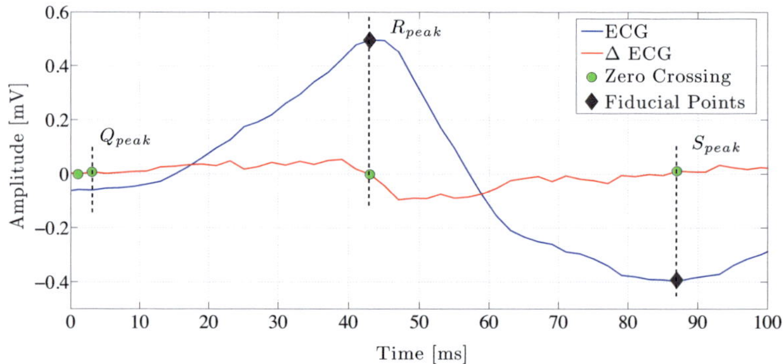

Fig. 7.7. QRS complex fiducial point detection after the delineation of the R peak. ECG signal in blue, differential signal of the ECG signal in red. Green dots represent the zero-crossings of the differential signal. Q and S peak can be detected by the zero-crossings of the differential signal.

7.2.2 Quality of the QRS delineation

The quality of the presented QRS delineation algorithm has been tested using the PVC-study. To quantitatively measure the quality, a correction rate was calculated:

$$CR_{QRS} = \left(1 - \frac{FN + FP}{TP}\right) \cdot 100\% . \tag{7.13}$$

In this equation, FP are erroneously detected beats, FN are all beats, which have not been detected and TP are all correct detections. By using a majority decision, the results of all leads were considered. It turned out that in case of the PVC-study with 290149 beats in the reference annotation, 289895 beats were correctly delineated ($TP = 289895$). 140 beats were wrongly detected ($FP = 140$), while 254 beats were missed ($FN = 254$). Additional values are presented together with the results of the beat classification in Table 7.2 in Section 7.4.

7.2.3 Discussion

This section showed how the delineation of the QRS complexes of a surface ECG can be performed using a wavelet based method as described in [37, 67]. One major problem in the field of wavelet based QRS delineation is finding a appropriate threshold. ECG signals can vary between different leads, subjects or ECG recorders. The decision of an adaptable threshold using a RMS-value is a first step in getting independent from the ECG source. However, the limitation of an RMS based threshold turns out by analysing ventricular ectopic beats. Fixed RMS based thresholds can lead to short RR intervals by detecting large amplitude T waves by mistake. One can think, this is not a problem, as RR intervals resulting from this misclassification are very short. Hence, just neglecting the second QRS complex of a very short RR interval can correct the mistake. In case of an ectopic beat, the last RR interval before the ectopic beat, named coupling interval, can be as short as the RR interval resulting from of the previously described misclassification. Hence, just neglecting one of the QRS complexes of a very short RR interval can also eliminate ectopic beats.

Thus, a reliable threshold has to be found, which assures the reliable detection of all QRS complexes including ectopic beats, but also avoids the detection of T waves as QRS complexes. The threshold has to be as low as possible (to detect all QRS complexes) and additionally as high as necessary (to detect no T wave by mistake). A satisfactory RMS based threshold for all ECG signals processed in this thesis was not found. Hence, the threshold had to be adapted to the underlying ECG signal. The introduced optimization of the threshold makes an automatic reliable QRS delineation algorithm possible.

The results of the presented QRS delineation algorithm is above average of other QRS delineation algorithms [60]. The correct rate of the validation process was best 99,8%. During the development of the threshold optimization, the algorithm was completely new implement and embedded into BSAT.

The following section is focused on the delineation of the T wave.

7.3 T wave Delineation

Fully automatic delineation of the T wave is still in focus of today's research. Especially T wave delineation in Holter ECG data is of increasing interest. The American Food and Drug Administration (FDA) calls for further research in automatic T_{end} detection for thorough QT studies [76]. The task of detecting the boundaries of the T wave in an ECG is difficult. The amplitude of the region of interest lies in the near of the baseline and this results in a low signal to noise ratio [59].

The approach of T wave delineation in this work is based on a correlation between the T wave regions in the recorded ECG signal and a set of templates, representing a subject specific T wave. The position of the best correlation is stored and the fiducial points of the template are projected onto the T wave. To be independent of the heart rate, a set of stretched and squeezed templates has to be generated, delineated and correlated to the signal.

T wave Filter

The T wave delineation process starts by an optimized filtering of the recorded ECG signal. As the T wave has its main frequency band in a region below 40 Hz, a high pass FFT filter with a cut-off frequency at 0.3 Hz and a low pass filter with a cut-off frequency at 40 Hz filters the ECG as described in Section 7.1.1 . To get rid of white noise in the T wave region, the signal is additionally filtered by a mean value filter with a window length of 30 ms. In further examination, the signals are named as $\boldsymbol{E}_{f1}$ for the FFT filtered signal and $\boldsymbol{E}_{sm}$ for the additionally smoothed signal. Figure 7.8 shows an example of an ECG and both filtered signals. It can be seen that the shape of the T wave is only slightly modified by the filtering, while the shape of the QRS complex is significantly changed.

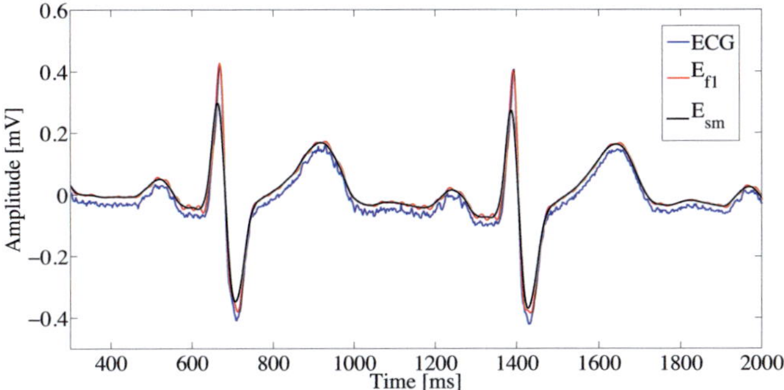

Fig. 7.8. Original ECG signal (blue line) and the output signals after denoising by a FFT high pass filter of 0.3 Hz and a low pass filter of 40 Hz (red line). Black line represents the filtered signal after a further mean value filter with a window length of 30 ms. A significant change in the P and T wave shape is not observed, while the shape of the QRS complex is clearly modified.

7.3.1 Template Generation

To find the T wave in the surface ECG, a correlation with a patient specific T wave template is made. Hence, the ECG signal has to be analysed to generate a template. Out of a huge number of T waves, which ideally shouldn't be influenced by artefacts, baseline wander or unusual morphology (caused by Premature Ventricular Contraction (PVC) for example), a patient specific template can be generated by calculating a mean wave.

Therefore, all regions in the ECG supposed to involve a T wave have to be found. Between two successive R peaks, detected in Section 7.2.1, a potential T wave region is supposed. The region starts at Reg_{Start}, lying 100 ms behind a detected R peak (R_i). R_i is the time stamp of the R peak of heart beat i. The end of the region is calculated by:

$$Reg_{End} = R_i + \frac{7}{2} \cdot (R_{i+1} - R_i) \tag{7.14}$$

If the distance between the end of the region and R_{i+1} is smaller than 150 ms, Reg_{End} is set to $R_{i+1} - 150\,ms$. Altogether, the length of the region has to be grater than 200 ms. If this is not satisfied, the heart beat is skipped. In every T wave region found, the maximum amplitude ($\hat{T}_{Max}$) is searched. The ECG signal between $\hat{T}_{Max} \pm 160$ ms is cut out and stored to the matrix $\boldsymbol{\Theta}_{Max}$. At this point of delineation, the information whether the T wave is positive or negative is not available. Thus, a second matrix $\boldsymbol{\Theta}_{Min}$ is generated with the signals around the minimum $\hat{T}_{Min}$ of every region.

Figure 7.9 shows Θ_{Max} for the example of a positive T wave. The high waves in the three dimensional plot are caused by T waves of PVCs. These and artefacts in the signal have to be detected, as they could lead to a faulty template. To de-

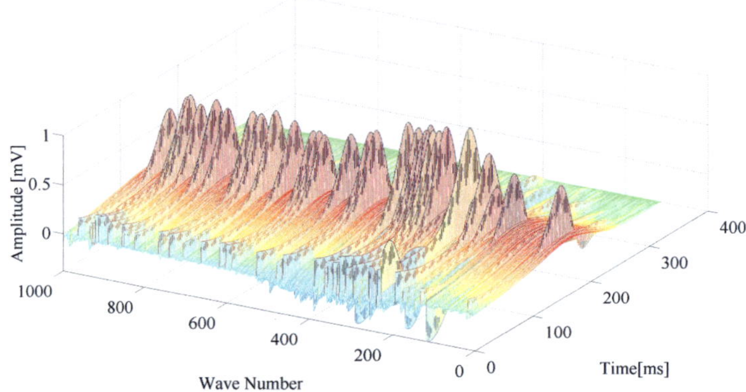

Fig. 7.9. 3D picture of the first 1000 extracted T wave regions used for template generation before outlier detection. Higher waves in the picture come from PVC and need to be thrown out by a further outlier detection. Only the first 1000 waves have been plotted for clarity reasons.

cide, whether the analysed T wave has a positive or a negative shape, templates are calculated out of both Θ_{Min} and Θ_{Max}. The mean of all supposed T wave regions represents the template. Two scenarios are possible:

- **Positive T wave**: The data in Θ_{Min} come from the region behind the T wave, as the lowest signal point is supposed to lie behind the T waves. Hence, the data in the matrix exists of baseline signal, maybe influenced by the end of the T waves in the left half. Data in Θ_{Max} involve all T wave regions of the ECG.
- **Negative T wave**: Data in Θ_{Min} involve the regions of the negative T waves of the ECG. As the highest signal point is supposed to lie behind the T wave, data in Θ_{Max} inhabits the regions behind the T wave. The data in the matrix exist of baseline signal, maybe left side influenced by the end of the T waves.

A decision whether the waves are positive or negative has to be made. A negative T wave has a significantly higher amount of negative values. The decision is made by the sum of negative values in Θ_{Min} (s_{Min}) and the sum of positive values in

Θ_{Max} (s_{Max}). It has turned out that a reliable measure for the determination of positive or negative T wave can be made by:

$$2 \cdot |s_{Max}| \geq |s_{Min}| \Rightarrow \text{ positive T wave} \tag{7.15}$$

$$2 \cdot |s_{Max}| < |s_{Min}| \Rightarrow \text{ negative T wave} \tag{7.16}$$

Depending on the decision, further analysis is done on Θ_{Min} or Θ_{Max}. It is named as Θ from now on.

Outlier Detection

Outliers in the T wave regions of the matrix Θ have to be detected. This is done in two steps. First, all waves having a strong shape difference to the main class are sorted out. For this purpose the sum of all sample values of each region is calculated as given in Equation 7.17.

$$\vartheta(i) = \frac{1}{N} \sum_{m=1}^{N} \Theta(i,m) \tag{7.17}$$

To distinguish the outliers from 'normal' waves, a threshold has to be found. The used threshold is based on the mean $(\overline{\vartheta})$ and the standard deviation (σ_ϑ) of the time series ϑ.

$$\overline{\vartheta} = \frac{1}{M} \sum_{i=1}^{M} \vartheta(i) \tag{7.18}$$

$$\sigma_\vartheta = \sqrt{\frac{1}{M-1} \sum_{i=1}^{M} \left(\vartheta(i) - \overline{\vartheta}\right)^2} \tag{7.19}$$

The time series calculated in 7.17 can be seen in Figure 7.10 as red dots. Based on the two values $\overline{\vartheta}$ and σ_ϑ, a interval I_ϑ, supposed to involve only 'normal' waves, can be found. Waves having a ϑ-value outside the interval are classified as outliers and sorted out.

$$I_\vartheta := \left[\overline{\vartheta} - \sigma_\vartheta, \overline{\vartheta} + \sigma_\vartheta\right] \tag{7.20}$$

Figure 7.11 shows the first 1000 T wave regions after the first outlier detection. The high waves coming from PVCs, as seen in Figure 7.9, have been eliminated from the ensemble. The variance between the waves is much smaller, but there are still some waves having a deformed morphology which need to be detected and

deleted. To detect the remaining outliers in the ensemble a Principal Component
Analysis (PCA) as introduced in Section 3.3, is performed on the rest of the data.
The PCA transforms data into a new space, where the axes are sorted by the data's
variance. T waves with a shape similar to the mean of the whole dataset will have
small scores. Waves with a deformed shape have higher scores. This explains the
need of the first outlier detection. Keeping the variance small makes the results
more reliable.

To break the group between 'normal' waves and outliers, a Hotelling's T^2-value is
performed (see Section 3.3). Equation 7.21 shows the computation.

$$\boldsymbol{T}^2 = \sum_{w=1}^{W} \frac{Score_w^2}{EV_w^2} \tag{7.21}$$

For every wave the value of T^2 is calculated. Waves having a value smaller than
the mean of all Hotelling's $\boldsymbol{T}^2$ values ($\bar{T}^2$) are classified as outliers and are sorted
out. Figure 7.12 shows the $\boldsymbol{T}^2$-values and the threshold. The residual T waves are
used to calculate the patient specific T wave template. The variance between the
remaining waves is very small, as shown in Figure 7.13.

Outlier detection using a PCA and Hoteling's T^2 measure might detect short or
long T waves as outliers. This effect is quested as the template should be generated

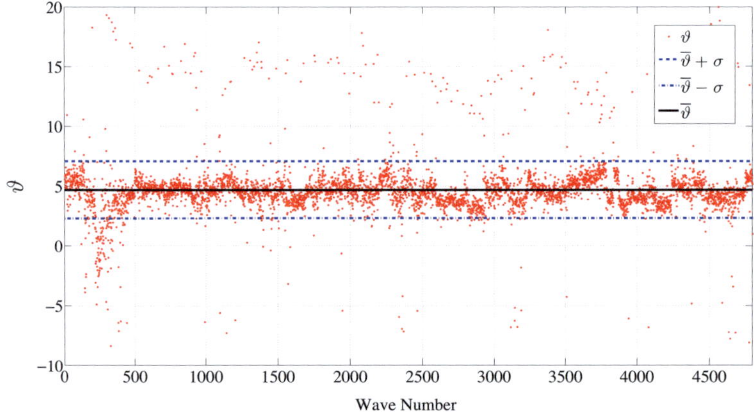

Fig. 7.10. Illustration of the first threshold for outlier detection in T wave template generation. Red dots
representing the surface area under the T wave. Interval of T waves supposed to be normal is marked by
dashed blue lines. Black line represents the median of all surface area values.

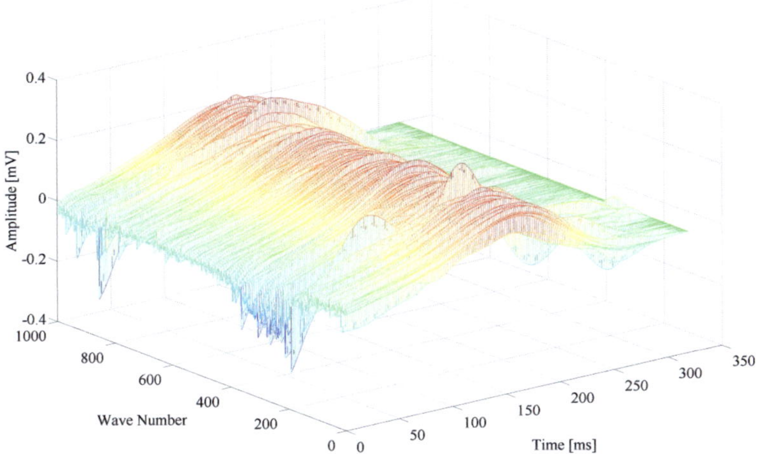

Fig. 7.11. Rest of first 1000 T wave regions after the first outlier detection step. High waves coming from PVC cannot be found any more. Some deformed T waves still remain in the candidate group.

from an ensemble of more or less equal T waves. Differences in the T wave coming from the variance of the heart rate will be considered by a squeezing and stretching of the template in a later step.

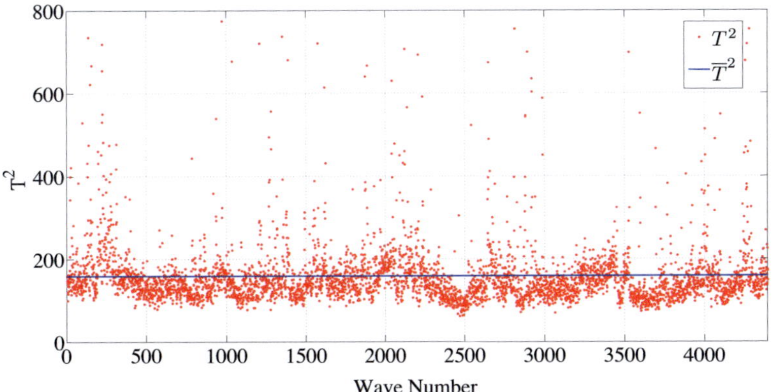

Fig. 7.12. Illustration of the PCA based thresholding for outlier detection in T wave template generation. The detection is made using the Hotelling's T^2-values (red dots). All T waves having a T^2-value smaller than the mean (blue line) can be used for template generation.

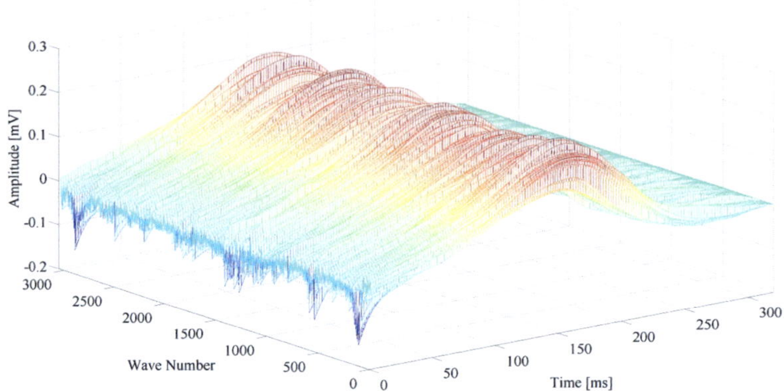

Fig. 7.13. 3D picture of all T waves remaining for T wave template generation after complete outlier detection. The variance of the T waves is small. Outliers from PVC or deformed T waves cannot be seen any more.

The patient specific morphology of the T wave is represented by the remaining T waves in $\boldsymbol{\Theta}, \widetilde{\boldsymbol{\Theta}}$. The template $\tilde{\boldsymbol{T}}$ is calculated as the mean of all remaining waves:

$$\tilde{\boldsymbol{T}}(i) = \frac{1}{\tilde{M}} \sum_{\tilde{m}=1}^{\tilde{M}} \widetilde{\boldsymbol{\Theta}}(\tilde{m}, i).$$

(7.22)

The so far generated template $\tilde{\boldsymbol{T}}$ can be seen in Figure 7.15 as blue line.

Template Delineation

The previously generated template needs to be analysed regarding its fiducial points. This involves the detection of the beginning T_{onset}, the peak T_{peak} and the end of the T wave T_{end}. The delineation process is part of this section. For the reason of clarity, the process is described for positive T waves, while it is analogue for negative ones. Figure 7.15 shows a T wave template with detected fiducial points. The delineation process starts by generating a differential signal of the template ($d\tilde{\boldsymbol{T}}$). In this signal the maximum and minimum is detected to get the inflexion points of $\tilde{\boldsymbol{T}}$. Maxima and minima of the signal $d\tilde{\boldsymbol{T}}$ represent the highest value of the positive and negative gradient of $\tilde{\boldsymbol{T}}$. Using the standard equation for a straight line, all necessary parameter for a tangent through each of the inflexion points can be calculated. Equation 7.23 and 7.25 show the mathematical description of the tangents for the inflexion points (ξ) at position $t_{\xi,1}$ and $t_{\xi,2}$.

$$\boldsymbol{v}_I(t) = d\tilde{\boldsymbol{T}}(t_{\xi,1}) \cdot t + b_1 \tag{7.23}$$

$$b_1 = \tilde{T}(t_{\xi,1}) - d\tilde{\boldsymbol{T}}(t_{\xi,1}) \cdot t_{\xi,1} \tag{7.24}$$

$$\boldsymbol{v}_{II}(t) = d\tilde{\boldsymbol{T}}(t_{\xi,2}) \cdot t + b_2 \tag{7.25}$$

$$b_2 = \tilde{T}(t_{\xi,2}) - d\tilde{\boldsymbol{T}}(t_{\xi,2}) \cdot t_{\xi,2} \tag{7.26}$$

The crossing of $\boldsymbol{v}_I$ with the baseline is marked as the onset of the T wave, T_{onset}. Similar, the end of the T wave, T_{end} is marked at the crossing of $\boldsymbol{v}_{II}$ and the baseline. The peak of the wave, T_{peak} is marked at the position of the maximum amplitude of the signal $\tilde{T}$. The detection of the fiducial points in case of a negative T wave is analogue.

Figure 7.14 shows an example of a T wave template $\tilde{T}$, the differential signal $d\tilde{T}$ and its corresponding fiducial points and tangents.

As the template of the T wave is generated out of regions in the ECG signal, the template might be longer as necessary. Hence, in the last step of delineation the template is shortened to a new template T_0, starting maximal 20 ms before T_{onset}

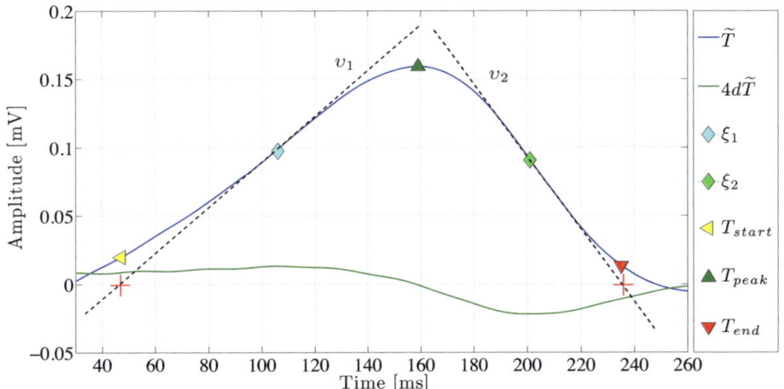

Fig. 7.14. Delineation of the T wave template. T wave (blue) and the differential signal (green). Diamonds represent the position of the T wave's inflexion points. Tangents through the inflexion points (black dashed lines) are used for T_{onset} and T_{end} localisation. Fiducial points are marked as coloured triangles.

and ending maximal $30\,\mathrm{ms}$ behind T_{end}. Figure 7.15 elucidates the shortening of the marked template.

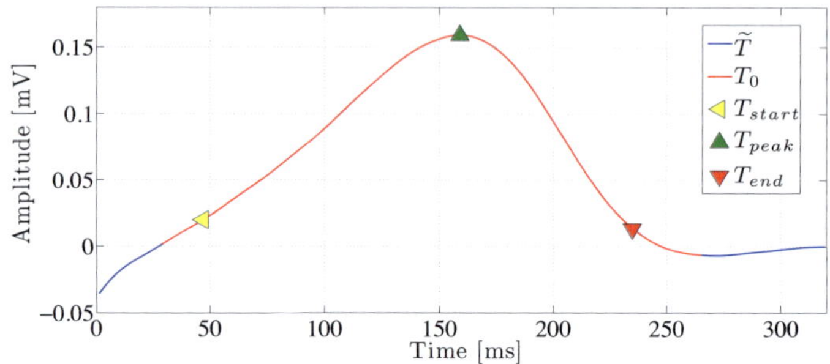

Fig. 7.15. Main T wave template (blue) and the used template curve (red) with its marked fiducial points after template delineation.

The use of the deflection points, tangents and zero crossings for finding T_{onset}, T_{peak} and T_{end} is quite common. In case other methods are preferred in future they can easily be applied to the template. Even a manual marking of T_{onset}, T_{peak} and T_{end} on the template might be used.

Set of T wave Templates

The generated, delineated and shortened template $\boldsymbol{T}_0$ is a representative subject specific T wave. However, the T wave morphology and duration is not necessarily constant for a whole recording. The T wave is dependent on the heart rate as described in Section 8.4. This is not yet considered in $\boldsymbol{T}_0$. The signal processing with its outlier detection leads to a T wave shape which represents only heart rates in a small range. To get a heart rate dependent template, the static signal $\boldsymbol{T}_0$ has to be transformed into a heart rate dependent function $\boldsymbol{T}_{RR}$.

In a recorded stress ECG, as shown in Figure 7.16, the observed T wave changes are mainly in height and length of the wave. The shape of the wave is nearly preserved. Altogether, the observed variance is comparatively small.

To satisfy both, minimal computational load and inclusion of changes of the T wave caused by heart rate adaptation during the ECG recording, the template is transferred into a set of nine templates. A transformation to a heart rate dependent template is not made. Hysteresis loops between QT and RR can often be observed in the ECG. A heart rate dependent T wave template needs to involve the subject

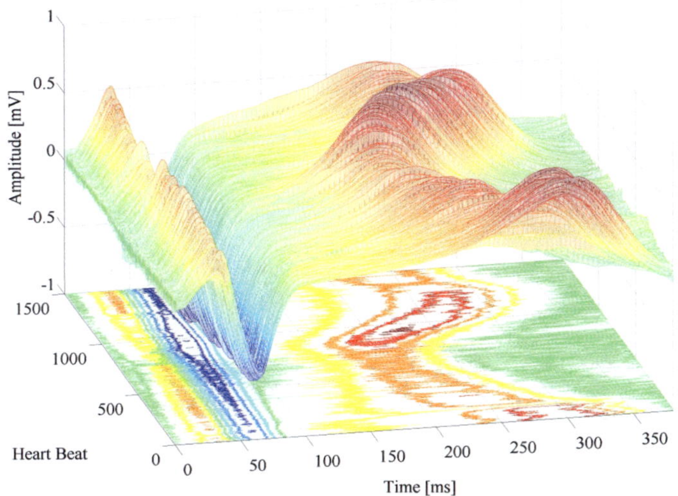

Fig. 7.16. Heart beats of a stress test ECG aligned at the R peaks in 3D axes. A change in QT time over heart beat number is observed. This is caused by an increasing heart rate due to the physical load. The main shape of the T wave is not significantly changed, while height and length of the T wave change. This can also be observed in the contour graph under the surface plot. ECG was recorded with *BSAT-1012*, amplitude value of the ECG recorder not calibrated.

specific QT/RR coupling. Modelling those relations to adapt the template to the signal is very difficult and not reliable. However, this is not necessary. The main template T_0 is squeezed and stretched to a set of nine templates (T_1 to T_9). The scale factors for squeezing and stretching are static: $sc := [0.8, 0.85, ... 1.15, 1.2]$. Figure 7.17 shows a set of templates and the fiducial points which have to be found for every template separately.

7.3.2 Correlation

Once a set of templates has been generated, the intrinsic fiducial point detection of the T wave can start. For this, the previously filtered ECG signal E_{f1} is used. Starting at the first detected R peak (R_1), a T wave region has to be defined to search for a T wave. The boundaries of the region are named as a and b. The region starts 50 ms behind an R peak, R_i. The end of the region is again calculated by Equation 7.14.

The QT interval is in any case shorter than 600 ms. To prevent computing time to

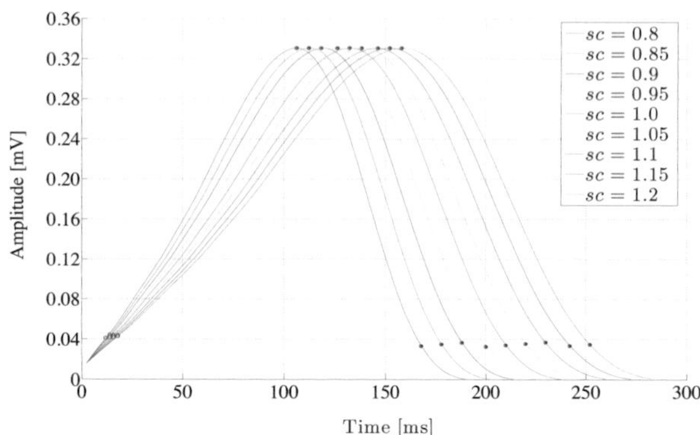

Fig. 7.17. Set of T wave templates. Subject specific template is squeezed and stretched between 0.8 and 1.2 to satisfy T wave variety coming out of heart rate adaptation. All nine templates are shown in this figure with corresponding fiducial points.

enlarge unnecessarily, b is set to $a + 600\,ms$ in case of a region length Δ_{ab} longer than 600 ms. On the other hand, a region has to be grater than the largest template T_9. Thus, the region bound b is set to at least a plus length of T_9 plus 50 ms. Altogether the region needs to end before the next detected R peak (R_{i+1}) and the heart rate is claimed to be smaller than 200 bpm. This limitation of maximum heart rate is made, as a RR interval smaller 300 ms and thus a T wave region smaller 250 ms, might lead to a faulty detection. Clinical ECG signals normally do not reach this limit and even if, this is caused by a tachycardia, which was not in focus of the signal processing in this research work. Stress test ECG recordings, as performed during this thesis, did not reach heart rates above 200 bpm.

Once a T wave region $\boldsymbol{\varepsilon}_i$ for heart beat i has been defined, a two step correlation process starts. During the correlation process two parameters have to be optimized. The position shift value sh and the template scale factor sc.

In the first step, the nine templates are cross-correlated with the region signal $\boldsymbol{\varepsilon}_i$. Equation 7.27 and 7.28 describe the cross correlation routine used in Matlab [77].

$$\hat{\boldsymbol{R}}_{xy}(m) = \sum_{n=0}^{N-m-1} \boldsymbol{x}(n+m)\boldsymbol{y}^*(n) \qquad m \geq 0 \qquad (7.27)$$

$$\hat{\boldsymbol{R}}_{xy}(m) = \hat{\boldsymbol{R}}^*_{yx}(-m) \qquad m < 0 \qquad (7.28)$$

The input vectors have to be of size N, otherwise the shorter vector is lengthened to N by adding zero values. In case of T wave delineation, templates are lengthened as well. As the template is shifted over the T wave region ε an output vector ψ results. Elements of ψ for the current heart beat and template i are described in Equation 7.29.

$$\psi_i(m) = R_{xy_i}(m-N), \qquad m = 1, 2, \cdots, N-1 \tag{7.29}$$

To get the position in the ECG signal the output vector has to be aligned correctly. Figure 7.18 shows the non zero part of ψ_i for all templates aligned to the corresponding ECG signal. A non scaled cross correlation was used. As the best position of every template is searched separately, it is not necessary to scale the output signal[1]. The position where a template T_i fits best to the T wave region ε_i is found by the maximum value of the output vector ψ_i.

After this step, a shift vector with optimal shift value for each template exists. Figure 7.18 shows an example which illustrates results of such a correlation process. In the upper part the ECG region of a T wave is drawn. In the lower part ψ is drawn for all nine templates. The maxima of the signals are marked by yellow dots. These represent the start positions of the templates (relating to the ECG signal) for their optimal fit to the T wave region.

Next, the scale factor sc, as the second free parameter in correlation process, has to be found. A linear correlation between the T wave region and each template at the position of the best shift sh_{best} is performed. To minimize the computational time the correlation and parameter optimization process is split into two steps. Linear or rank correlation which are comparable against each other are time consuming. Instead of testing all positions and templates, the maximum of the less time consuming non scaled cross-correlation is searched and then the templates are compared against each other at the optimal fit position, using the linear correlation. The algorithm computes Pearson's linear correlation coefficient to compare the templates. In equation 7.30 the templates and the corresponding samples of the T wave region are represented by the vectors x and y. The variables $\bar{x}$ and $\bar{y}$ represent the mean of x and y.

---

[1] The computational time is significantly increased for an equal but scaled correlation sequence.

$$r_{xy} = \frac{\sum\limits_{l=1}^{n} (\boldsymbol{x}(l) - \bar{x})(\boldsymbol{y}(l) - \bar{y})}{\sqrt{\sum\limits_{l=1}^{n} (\boldsymbol{x}(l) - \bar{x})^2 \sum\limits_{l=1}^{n} (\boldsymbol{y}(l) - \bar{y})^2}} \tag{7.30}$$

The template $\boldsymbol{T}_{best}$, with the maximum linear correlation coefficient r_{xy}, fits best to the T wave region. The scale factor is set to the scale factor of this template.

To delineate the T wave of the current heart beat i the fiducial points of $\boldsymbol{T}_{best}$ can be projected onto the ECG signal. As the shift of the template and the fiducial points of the template are known, the fiducial points of the T wave can be calculated by:

$$T_{onset} = R_i + sh_{best} + T_{onset,best}, \tag{7.31}$$

$$T_{peak} = R_i + sh_{best} + T_{peak,best}, \tag{7.32}$$

$$T_{end} = R_i + sh_{best} + T_{end,best}. \tag{7.33}$$

The value of sh_{best} gives the position of the beginning of the best fitting template at the position with the highest correlation in the total ECG segment. sc_{best} and sh_{best} are valid for the specific heart beat they have been found for.

Large signal amplitudes have a higher impact on the correlation coefficient than small amplitudes. Thus, a delineation of the T wave is focused on the peak of the wave and not on the end of the T wave. In clinical routine, mostly the end of the T wave is requested. So, the stronger weighting of the T wave peak in the correlation is a drawback at the expense of the T_{end} delineation accuracy. To balance this drawback a refinement is performed.

Refinement

The refinement is done by minimizing the difference D between the template $\boldsymbol{T}_{best}$ and the corresponding ECG sequence in a refinement interval of $t_{ref} = 10\,ms$ around the position determined previously. To focus on the end of the T wave, a multiplication vector $\boldsymbol{w}_i$ of the length of $\boldsymbol{T}_{best}$ is defined.

$$\min_{k} D(k) = \sum_{m=1}^{M} \boldsymbol{w}_i(m) \left(\boldsymbol{E}_{sm}\left(sh_{best} - \frac{t_{ref}}{2} + m\right) \cdot \boldsymbol{T}_{best}(m) \right)^2. \tag{7.34}$$

Vector $\boldsymbol{w}_i$ weights the differences between $\boldsymbol{T}_{best}$ and the ECG signal. All values before T_{peak} (of $\boldsymbol{T}_{best}$) are multiplied by 0.5, while all values at, or behind T_{peak} are weighted with 1. The computation of $\boldsymbol{D}$ is done for every sample point in the

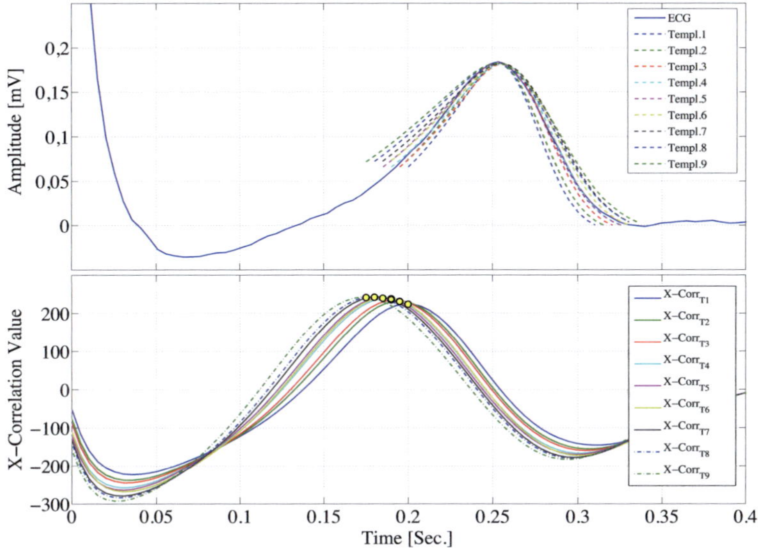

Fig. 7.18. Correlation of the whole template set with a T wave region. Upper axes: ECG signal with all templates positioned for best correlation. Lower Axes: cross correlation vector of all templates aligned to the relating position in the ECG signal. Yellow dots mark the best correlation position for each template separately.

refinement interval. Thus, for a signal with a sample rate of $f_s = 1000\,\text{Hz}$, k is in the interval between 1 and 10. The lowest value in $\boldsymbol{D}$ represents the refinement shift of the T wave. This shift is added to the fiducial points of the current T wave. The refinement is the last computation step for T wave delineation of heart beat i. The fiducial point detection of the i+1-*th* T wave is started and done in the same manner.

Once the T wave delineation process has been done for all detected heart beats in the ECG signal, the results are stored in a fiducial point table. This table is the basis for all further signal processing and time series generation as introduced in Section 7.5. The ECG delineation process is completed at this point.

7.3.3 Results of the Delineation

The T wave delineation process described in this chapter was developed within this research work. During the delineation process, every heart beat of the ECG is marked. A high reliability of the algorithm is reached because of the subject specific template generation. This enables an extraction of clinically relevant information out of the ECG signal, for example by analysing time series involving information of changes in wave morphology. The usage of mean values of ten second intervals as commonly performed for QT interval analysis [78] is neither necessarily nor recommended.

The developed T wave delineation with its T_{end} detection was tested in several manners. A first validation was done on the MIT QT-Database [79]. This database involves 105 short term ECG recordings having a manual expert annotation of about 15 heart beats, per tape. In this work 92 recordings have been tested. The other records have been excluded for technical reasons. In some cases the ECG data could not be read in, in other cases expert annotation was not available or readable. In total 3015 heart beats of 92 ECG recordings have been compared to the experts annotations.

Annotation Comparison

A golden standard definition of the T_{end} position in the ECG does not exist. The author choose the tangent method [69] as one of the most common methods for T_{end} delineation. However, the absence of a golden standard leads to differences in T_{end} detection. Those differences lead to a constant offset error. The offset makes the comparison more complex, but not impossible. To be able to compare the expert annotation and the annotation of the automatic T wave delineation algorithm, a range of permitted offsets have to be defined. All T_{end}-values having a difference to the corresponding expert annotation smaller than O_{max} are counted as correctly detected, (d_+). All T_{end}-values having a lager difference to the corresponding expert annotation are counted as falsely detected (d_-). A detection rate can be calculated by Equation 7.35. In this, L is the total number of compared waves. As for every detected R peak a T wave is found by the best correlation in a T wave region, the calculation of sensitivity or specificity is not reasonable at this point. These values are presented for the R peak detection in Section 7.4.5.

$$D_{Rate} = \frac{d_+}{L} \cdot 100\% \tag{7.35}$$

Results of Annotation Comparison

Table 7.1 and Figure 7.19 show the results in dependency of the permitted offset. The mean error and the standard deviation of the detected waves are shown as well. It can be observed that the detection rate increases by an increase of the allowed offset. The mean error increases as well, mostly until O_{max} reaches 80 ms. For $O_{max} \geq 80ms$ the mean error $\bar{e}$ remains approximately constant. As the mean error is negative, the experts annotation is delayed in the signal compared to the automatic annotation.

The standard deviation $\bar{\sigma}$ gives an overview on the variance of the differences. σ increases with the detection rate. For a permitted offset of more than 100 ms $\bar{\sigma}$ is less increasing. The increase of the standard deviation is a corollary of the increase of the permitted offset. In summary, a permitted O_{max}-value grater than 100 ms does not significantly increase the values of $\bar{\sigma}$ and $\bar{e}$. A maximum permitted error of 100 ms leads to a detection rate of 91.19%. The results of every single record can be found in the Appendix. Table A.1 shows the results for $O_{Max} = 100ms$ for all records.

Alternative Validation

Validation of fiducial point detection has to be made in comparison to manual expert annotation. However, this is a problem as also the manual annotation of ECG

Table 7.1. Results from the MIT BIH QT Database for the detection of T_{end} in dependence of the permitted offset O_{max}. Comparison between experts manual annotations and the fully automatic delineation. Detection rate D_{Rate}, mean error $\bar{e}$ and mean standard deviation $\bar{\sigma}$ are shown.

O_{Max} [ms]	D_{Rate} [%]	$\bar{e}$ [ms]	$\bar{\sigma}$ [ms]
10	33.23	-1.16	4.89
20	59.43	-4.99	8.40
30	69.23	-7.28	9.89
40	77.68	-9.55	11.83
50	80.30	-11.52	12.34
60	84.05	-12.59	13.50
70	85.95	-13.22	14.20
80	88.48	-14.02	14.84
90	89.75	-14.14	15.28
100	91.19	-14.14	15.86
110	91.93	-15.60	16.21
120	93.40	-14.45	16.72
130	94.20	-14.37	16.84
140	95.88	-14.43	17.15
150	96.75	-14.16	17.34

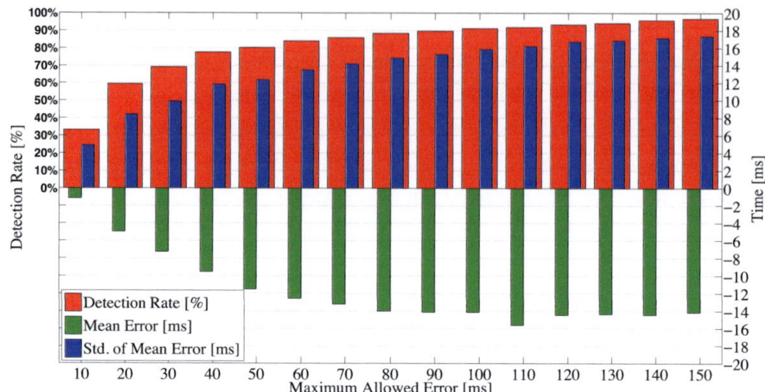

Fig. 7.19. Detection rate in dependency of allowed error

signals is not 100% exact. All errors in manual annotations lead to a false delineation result of the algorithm. This requires a validation in the scenario for which the algorithm has been developed.

This work focuses on the T wave delineation of Holter ECG data to detect drug induced repolarisation changes. To validate the algorithm in a further step, it was used to detect the T wave in ECG data of the THEW tQT database I and II. An introduction to the databases can be found in Section 5.3. The QT interval was extracted and heart rate corrected by the formula of Bazett [80]. A QT prolongation of about 12 ms under the influence of Moxifloxacin has been observed. Khawaja et al. in [78] and Couderc et al. in [81] came to similar results of QT prolongation by analysing these THEW ECG studies. Figure 7.20 shows a boxplot[2] of the QT prolongation for the THEW tQT_{II} study. The introduced fully automatic delineation algorithm can be used to detect even small drug induced QT prolongation in Holter ECG data. A detailed explanation of drug induced QT prolongation and further results are presented in Chapter 8, *'Drug Induced Repolarisation Changes'*.

7.3.4 Discussion

The introduced T wave delineation algorithm showed good performance and reliability. The detection of the end of the T wave was possible, even under bad conditions. A detection rate of 91.15% has been achieved for the MIT QT database

---

[2] The information given by the boxplots used in this work is specified in Section 8.2.3

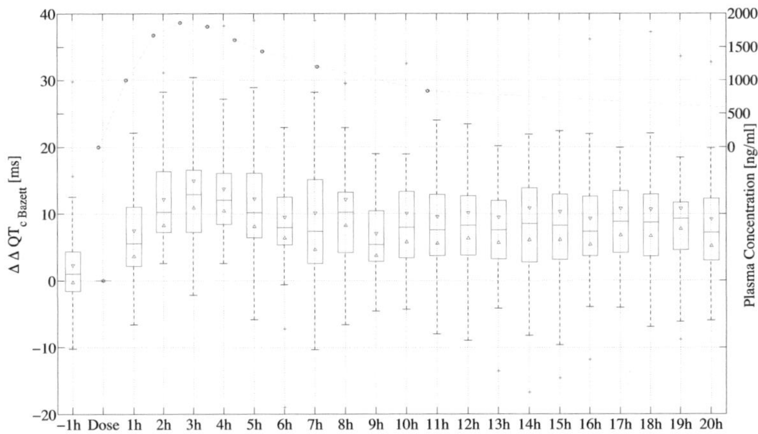

Fig. 7.20. Boxplot of QT prolongation of THEW tQT_{II} study. QTc is calculated using the correction formula of Bazett. Interval length is 60 minutes. 57 subjects have been considered. Plasma concentration of the drug Moxifloxacin is plotted in green. A high correlation of the plasma concentration of Moxifloxacin and the QT-prolongation can be observed. $\Delta\Delta QT_c$ is the difference between two recordings with once placebo and once Moxifloxacin administered. Day to day variance is eliminated by setting QT_c to zeros predose.

for a maximum permitted error of 100 ms. Moxifloxacin induced QT prolongation of 12-14 ms could be detected by analysing the QT interval extracted from ECG's delineated by the method.

As a T wave is found for every detected heart beat, the T wave detection rate is always 100%. To get a meaningful value for the detection quality, a maximum permitted error has to be defined. As there is no rule for the value of a maximum allowed error, the results have been computed for different values between 10 ms and 150 ms. Calculation of sensitivity and specificity is not meaningful. As previously reported, the algorithm finds a T wave for every heart beat detected by the QRS delineation. Hence, sensitivity and specificity are strongly depended on the QRS delineation. Results of QRS detection and classification are presented in Section 7.4.5. To analyse the quality of the T wave-delineation algorithm separately, the detection rate as described before is the best choice.

The validation of the delineation algorithm via a comparison to manual annotations has a major drawback. The manual experts annotations can have a variance themselves. Noise, artefacts or baseline wander can cause this variance. Further more, differences between ECG recordings within a study caused by different wave morphologies in different leads might lead to a variance in manual anno-

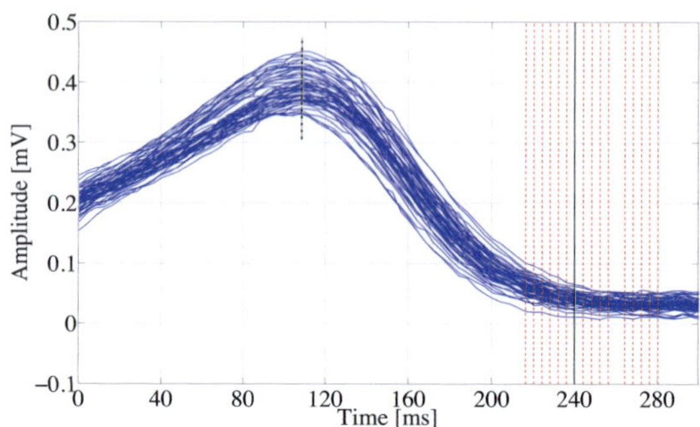

Fig. 7.21. Differences of manual annotations. Wave ave been aligned to their peaks. The red dashed lines show the positions of the T_{end} annotations.

tations. Within a group of experts an individually related variance is also most likely. Figure 7.21 shows an example of a manual annotation of 30 T waves made by one expert within a single ECG recording. The T waves have been aligned on T_{peak}. The standard deviation of the T_{end} annotations is very high. Interestingly, the standard deviation of the manual annotation after alignment is 19.1ms, and thus larger than the maximum mean $\overline{\sigma}$ seen in Table 7.1 . Of course, a direct comparison of both values should not be made, but it illustrates the limitations of this validation method.

Besides the validation using manual annotations, clinical data with a known information can help to proof the quality of a delineation algorithm. The drug Moxifloxacin is known to prolong the QT interval. If the algorithm reliably detects small drug induced QT prolongation, a high delineation quality can be assumed.

The THEW tQT_{II} study is very powerful for the validation of automatic delineation algorithms. On the one hand, the drug Moxifloxacin is known for its impact on the QT interval. On the other hand, a Moxifloxacin blood plasma concentration curve for all recordings is available. A strong correlation between the blood plasma concentration and the QT prolongation measured by the algorithm makes the validation results more reliable.

Indeed, QT_c prolongation due to Moxifloxacin can be observed in $\Delta\Delta$-plots generated from the delineated ECG. As observed in Figure 7.20, $\Delta\Delta QT_c$ increases post dose, until about three hours after the drug was administered. A maximum QT_c prolongation of about 12 ms is ascertained. A strong correlation to the plasma concentration curve is observed. The boxplot shows the results of 57 subjects. QT_c values have been averaged in intervals of 1 hour.

A combination from both, comparison of manual expert annotation and testing clinical issues in a well known dataset leads to a validation process which provides useful information on the quality of the delineation algorithm. The validation process for this T wave delineation algorithm turned out high reliability and high quality for the correlation based method. In *'Comparison of three T-Wave Delineation Algorithms based on Wavelet Filterbank, Correlation and PCA'* [5], the author of this work implemented and tested two additional T_{end}-delineation algorithms. It turned out that the correlation based method works most reliable and leads to small delineation errors.

As any method, also T wave delineation by the correlation based algorithm has some limitations. The introduced method is based on a subject specific template generated from the recorded ECG. The generation of the template assumes a high number of 'normal' T waves in the ECG data, as they can be found in healthy subjects. Subjects suffering from heart diseases which affect the morphology of the T wave and change it significantly during recording time might be a challenge for the delineation algorithm. Significant changes of the wave morphology can lead to a wrong template and consequently to erroneous T wave delineation. As a rule of thumb, the ECG should involve at least 70% 'normal' heart beats.

The delineation of the template has been optimized for positive and negative T waves. In all datasets used in this work, leads with either positive or negative T waves have been found. In some ECG recordings bi-phasic T waves might be found. In such case, a delineation by the algorithm is possible, even though the delineation might be inexact. The T wave might be seen as positive or negative, which leads to a delineation of either the first or the second half of the wave. An update of the template delineation routine with an additional determination of bi-phasic T waves could be possible. Biphasic T waves have not been considered and thus a reliable delineation of ECG recordings inhabiting those waves is not recommended yet. In an update, bipasic T waves could be considered and the template delineation routine could be augmented. The correlation routine of the algorithm needs not to be changed for this purpose.

Delineation of ECG data with T wave alternans has to be discussed in this context. T wave alternans, found in the early years of the 20th century was considered as a seldom phenomena [82]. This T wave alternans is related to a significant morphology variation which could also lead to problems in the T wave delineation using the correlation based method. From the 1980s on, a computational analysis of T wave alternans was possible and the phenomena of nonvisible (microvolt-level) alternans was studied [83]. This alternans should not affect the delineation process, as the variance between the beats is small. The template should not significantly change and a reliable correlation process will be possible. A variance of the maximum amplitude of the T wave is not problematic, as the best correlation in a T wave region is not changed. An alternans in the duration of the T wave can be compensated by the discriminative scaled templates. In summary, microvolt T wave alternas does not affect the delineation process. In the seldom case of significant T wave alternas, a method based on one single template in general is not recommended.

7.4 Beat Classification using a Support Vector Machine

The QRS complex in the surface ECG indicates ventricular depolarisation. This important information can be used to analyse the heart rhythm of the patient. Additional information whether e.g. a normal heart beat, a Premature Ventricular Contraction (PVC)[3] or an Premature Atrial Contraction (PAC)[4] was detected is not available. A further analysis is necessary to classify all detected heart beats. In this thesis three classes of beats have been distinguished:

- Normal heart beats coming from the sinus node in normal rhythm. These beats are labelled as R_N.
- Premature atrial contraction, arise in the atrium, but too early for the normal rhythm. These beats are labelled as R_A.
- Premature Ventricular Contraction, arise in the ventricles. They have different morphology and appear before the next normal beat. These beats are labelled as R_V.

Figure 7.22 shows an ECG signal with a PAC, while Figure 7.23 shows a PVC in the same ECG recording. To distinguish between normal beats, PVC and PAC, rhythmical and morphological parameters of each QRS complex are computed and

---

[3] PVC is also named: extrasystole, ventricular premature beat or ventricular premature contraction.

[4] PAC is also named: atrial premature complexes or atrial premature beats

analysed. A description of these parameters is provided in Section 7.4.1 and 7.4.2. The classification of the beats is done with a Support Vector Machine (SVM). The performance of the QRS detection and beat classification was validated on the PVC-study introduced in Section 5.7. The results are presented in Section 7.4.5.

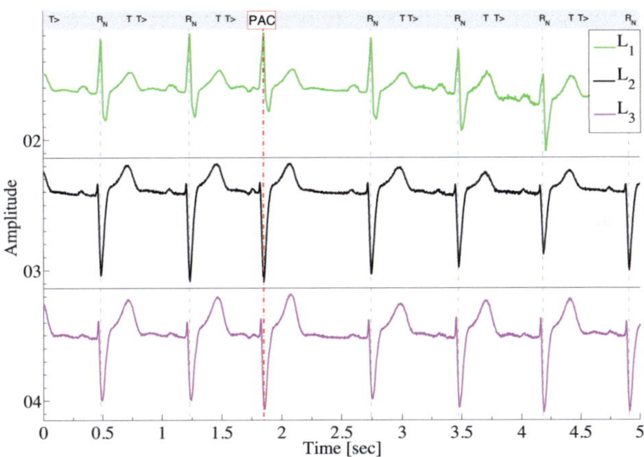

Fig. 7.22. ECG signal segment of PVC-study subject 1 with PAC. RR interval before PAC is shorter than previous RR intervals. RR interval after PAC is longer than the following RR intervals. The morphology of the PAC is equal to those of normal beats.

7.4.1 Rhythmical Descriptors of the QRS Complexes

To distinguish normal beats from PVC or PAC, rhythmical features of the RR intervals are analysed. In Figure 7.22 it can be seen that the RR interval before the PAC is shorter than normal, while the following RR interval is longer than normal. Similar observations can be made in Figure 7.23 for a PVC.

It is obvious to analyse rhythmical features to distinguish PVCs from normal heart beats. The presented classifier uses 20 different rhythmical features in the classification process. A detailed description of all rhythmical features is presented in the supervised Student Research Project of Gustavo Lenis [4]. Here, a short overview over the most important rhythmical features is given.

Equation 7.36 and 7.37 show the relationship of the R peak (as defined in Section 7.2.1) of a heart beat i to the corresponding RR interval (RR_i) and the heart rate

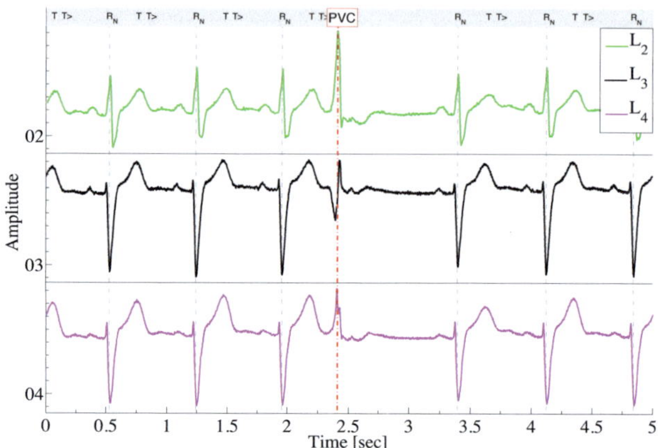

Fig. 7.23. ECG signal segment of PVC-study subject 1 with PVC. RR interval before PVC is shorter than previous RR intervals. RR interval after PVC is longer than the following RR intervals. The morphology of the PVC is clearly different to those of the normal beats.

(HR_i).

$$RR_i = R_{i+1} - R_i \tag{7.36}$$

$$HR_i = \frac{60}{RR_{i-1}} \tag{7.37}$$

As already mentioned, RR intervals before an ectopic beat are shorter than RR intervals of normal beats, while the subsequent beat has a longer RR interval. Hence, the first group of rhythmical features is based on the relationship between RR_i and RR_{i-1}. So-called Poincaré plots are common to elucidate self-similarity in a process. Figure 7.24(a) shows a Poincaré plot of RR_i vs. RR_{i-1}. During normal sinus rhythm, RR interval lengths are supposed to lie in a cloud around the bisection of the first quadrant, while PAC and PVC are supposed to lie above the bisection. Related to Equation 7.36, for every feature generated from RR interval values a corresponding feature generated from the values of the heart rate exists. The heart rate based features are linear independent from the RR interval based features and they should bring new information to the classification process.

The first feature introduced, is based on the difference of two successive RR in-

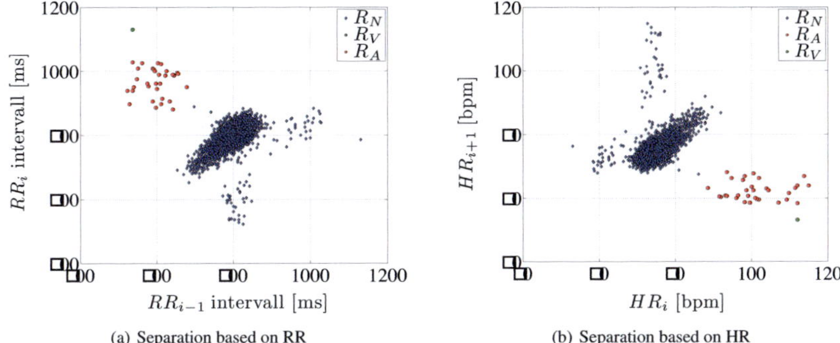

Fig. 7.24. Poincaré plots of a) RR_i vs. RR_{i-1} and b) HR_{i+1} vs. HR_i of ECG signal 100 of the MIT-BIH arrhythmia database. In a) the ectopic beats lie above the bisection. Beats below the bisection with a distance to the main cloud come from beats previous or subsequent an ectopic beat (the coupling interval or the compensatory pause influences the position in the plot). In b) the results are mirrored.

tervals. Normal beats will have small values around zero, while ectopic beats will have larger values. An example of the difference between two successive RR intervals can be seen in Figure 7.25. Equations 7.38 and 7.39 show the calculation.

$$\Delta RR_i = RR_i - RR_{i-1} \tag{7.38}$$

$$\Delta HR_i = HR_i - HR_{i+1} \tag{7.39}$$

To scale down the cloud of normal beats, also the quotient of two successive RR intervals and heart rates is calculated. Figure 7.24 elucidates this principle in general. The inter-beat difference is small for normal beats and thus the quotient of the length of two successive RR intervals is around one. Equation 7.40 and 7.41 show the calculation of two features using this fact:

$$qRR_i = \frac{RR_i}{RR_{i-1}} \tag{7.40}$$

$$qHR_i = \frac{HR_{i+1}}{HR_i}. \tag{7.41}$$

To increase the difference between normal- and ectopic beats, an additional ratio is calculated, considering the RR intervals and heart rates of the last two beats.

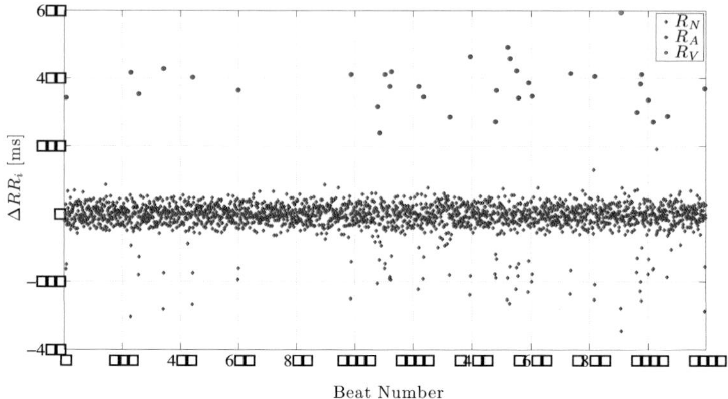

Fig. 7.25. Difference between two successive RR intervals (ΔRR) over time. Differences of normal beats lie near to zero, while PVC and PAC have significantly increased values.

$$QRR_i = \frac{qRR_i}{qRR_{i-1}} = \frac{RR_i \cdot RR_{i-2}}{RR_{i-1}^2} \tag{7.42}$$

$$QHR_i = \frac{qHR_i}{qHR_{i-1}} = \frac{HR_{i+1} \cdot RR_{i-1}}{HR_i^2} \tag{7.43}$$

As the heart rate varies throughout the day, a mean heart rate and a mean RR interval in a window of 11 beats is calculated to get a measure of the current heart rate in the region of an ectopic beat. Equation 7.44 and 7.45 show the computation. The value of the current heart beat i is neglected as it may be an ectopic beat.

$$\overline{RR}_i = \frac{1}{10} \sum_{j=i-5, j\neq 5}^{i+5} RR_i \tag{7.44}$$

$$\overline{HR}_i = \frac{1}{10} \sum_{j=i-5, j\neq 5}^{i+5} HR_i \tag{7.45}$$

Using $\overline{RR}$ and $\overline{HR}$, additional descriptors can be generated. The following equations give an impression on some of the descriptors:

$$q\overline{RR}_i = \frac{\overline{RR}_i}{RR_i} \tag{7.46}$$

$$q\overline{HR}_i = \frac{\overline{HR}_i}{HR_i} \tag{7.47}$$

$$\Delta q\overline{RR}_i = qRR_i - \frac{1}{10} \sum_{j=i-5, j\neq5}^{i+5} qRR_j \tag{7.48}$$

$$\Delta q\overline{HR}_i = \frac{1}{10} \sum_{j=i-5, j\neq5}^{i+5} qHR_j - qHR_i \tag{7.49}$$

$$Q\overline{q}RR_i = \frac{qRR_i}{\frac{1}{10} \sum\limits_{j=i-5, j\neq5}^{i+5} qRR_j} \tag{7.50}$$

$$Q\overline{q}HR_i = \frac{qHR_i}{\frac{1}{10} \sum\limits_{j=i-5, j\neq5}^{i+5} qHR_j}. \tag{7.51}$$

In total, 20 rhythmical descriptors are generated. These descriptors are labelled RP_1 to RP_{20} from now on. The rhythmical parameters are used in the SVM to determine to which of the three classes a beat belongs. In Figure 7.24 the classification seems to be very easy. All three types of beats can be clearly distinguished. However, the differences in the pattern of the three classes are not always clearly distinguishable. Figure 7.26 shows an ECG signal of the MIT-BIH arrhythmia database. In this case the classes overlap, especially for both ectopic beat classes. The PAC class and normal beats still have different rhythmical properties in Figure 7.26. For some of the PVCs, the rhythmical properties are quite similar to those of normal beats. However, QRS complexes of PVCs have a different morphology compared to normal beats and PACs. This can be seen in Figure 7.23. Thus it can be concluded that an analysis of the QRS morphology delivers additional information that might ultimately help to improve the classification of the different beats. In the next sections a short introduction on morphological features for beat classification is given.

7.4.2 Morphological Descriptors of the QRS Complexes

Besides the different rhythmical properties that help to distinguish between normal- and ectopic beats, the morphology of the QRS complexes of PVC differs as well. PVCs that have very similar rhythmical features compared to normal beats, can be

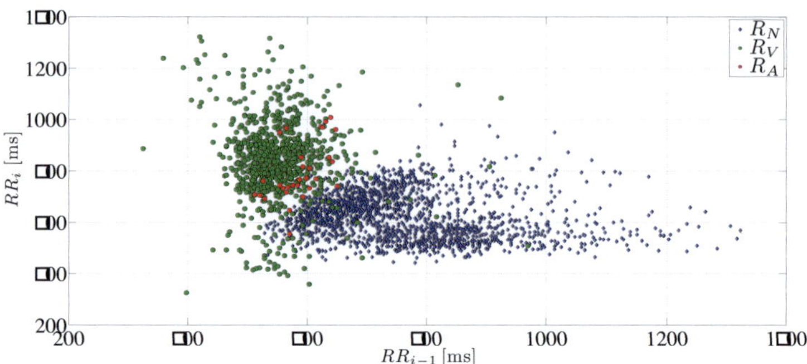

Fig. 7.26. Overlap of beat classes in a Poincaré plot showing RR_i vs. RR_{i-1} of an ECG signal from MIT-BIH arrhythmia database. Normal beats (blue dots) cannot be separated from ectopic beats (green and red dots). The classes of PAC and PVC also overlap.

correctly classified by taking morphological descriptors into account. As the morphology of the QRS complex of a PAC is similar to the QRS complex of a normal beat, morphological descriptors can be used to separate PVCs from both, normal beats and PACs.

In this section the morphological QRS descriptors are introduced. The first morphological parameters are based on the amplitude of the QRS complex:

$$MP_1 = max\{\boldsymbol{s}\} \tag{7.52}$$

$$MP_2 = min\{\boldsymbol{s}\} \tag{7.53}$$

where $\boldsymbol{s}$ represents a vector of amplitudes of a QRS complex. Generally speaking, absolute amplitude based values in biosignal processing can be unreliable. However, the maximum and minimum amplitude of normal beats and PVCs can be very different. A separation of the beat classes is often possible. To take the slope of the QRS complexes into account, descriptors based on minima and maxima of the first and second derivative of $\boldsymbol{s}$ are used. The minima and maxima of those signals are used.

$$MP_3 = max\{\boldsymbol{s}(i) - \boldsymbol{s}(i-1)\} \tag{7.54}$$

$$MP_4 = min\{\boldsymbol{s}(i) - \boldsymbol{s}(i-1)\} \tag{7.55}$$

$$MP_5 = max\{[\boldsymbol{s}(i) - \boldsymbol{s}(i-1)] - [\boldsymbol{s}(i+1) - \boldsymbol{s}(i)]\} \tag{7.56}$$

$$MP_6 = min\{[\boldsymbol{s}(i) - \boldsymbol{s}(i-1)] - [\boldsymbol{s}(i+1) - \boldsymbol{s}(i)]\} \tag{7.57}$$

In addition to amplitude based descriptors, area based descriptors are used.

$$MP_7 = \frac{1}{f_s} \cdot \sum_{j=1}^{N-1} \frac{s(j) + s(j+1)}{2} \qquad (7.58)$$

in this equation f_s represents the sample frequency. To calculate the positive and negative area under the QRS complex two new signals are introduced:

$$s_p(j) = \begin{cases} s(j), & \text{if } s(j) > 0 \\ 0, & \text{else} \end{cases} \qquad (7.59)$$

$$s_n(j) = \begin{cases} s(j), & \text{if } s(j) < 0 \\ 0, & \text{else} \end{cases} \qquad (7.60)$$

Using s_p and s_n together with Equation 7.58, the positive and negative area of the QRS complex is calculated as MP_8 and MP_9, respectively. The last area based feature introduced here uses the square of the signal amplitude:

$$MP_7 = \frac{1}{f_s} \cdot \sum_{j=1}^{N-1} \frac{s^2(j) + s^2(j+1)}{2} \qquad (7.61)$$

Differences in the amplitude and area under the curve are based on the fact that the signal spectrum of a PVC is different to the spectrum of a normal beat or PAC. Hence some morphological descriptors are based on the FFT of the QRS complex. The absolute value of the spectrum of the signal

$$|FFT\{s(j)\}| = |S(k)| \qquad (7.62)$$

is divided into three intervals:

- F_1 with the frequencies interval between 5 Hz and 10 Hz
- F_2 with the frequencies interval between 10 Hz and 20 Hz
- F_3 with the frequencies interval between 20 Hz and 50 Hz

The spectrum of PVCs is expected to have increased values compared to normal beats and PACs in the interval F_1. For both intervals F_2 and F_3 the values are expected to be decreased for PVCs. Three parameters are generated:

$$MP_{11} = max \; \{|S(f)|\} \, , \, f \in F_1 \tag{7.63}$$

$$MP_{12} = mean \; \{|S(f)|\} \, , \, f \in F_2 \tag{7.64}$$

$$MP_{13} = max \; \{|S(f)|\} \, , \, f \in F_3 \tag{7.65}$$

To describe the morphology of a QRS complex, Hermite basis functions can be used. The parameters a_1 to a_3 of the Hermite basis functions are optimized for an optimal fit of the corresponding QRS complex. Different morphologies of the QRS complexes can be distinguished by comparing the parameters of the Hermite basis functions. In this thesis a representation of Hermite functions as introduced in Section 3.4 is used to approximate the QRS complexes.

$$s(j) \approx a_1 \cdot h_1 \left(\frac{j-\tau}{\sigma} \right) + a_2 \cdot h_2 \left(\frac{j-\tau}{\sigma} \right) + a_3 \cdot h_3 \left(\frac{j-\tau}{\sigma} \right) \tag{7.66}$$

In this equation τ is the centre of mass and σ is the variance. They are chosen in dependency of the centre of mass and the variance of the actual QRS complex. h_1 to h_3 represent the Hermite functions of order 1 to 3. The parameters a_1 to a_3 are used as MP_{18} to MP_{20}. Figure 7.27(a) shows a three dimensional scatter plot of the descriptors resulting from the approximation.

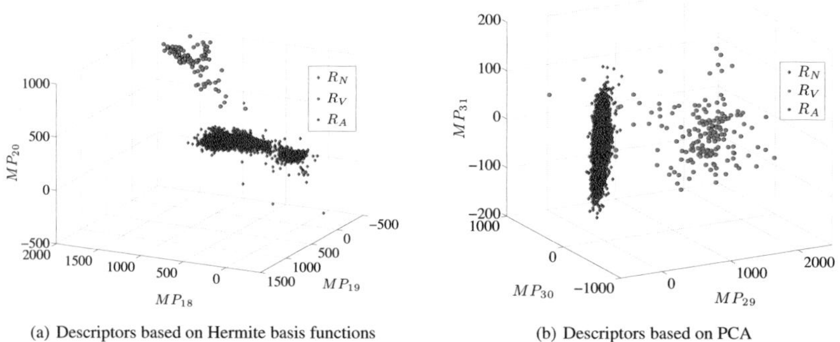

(a) Descriptors based on Hermite basis functions   (b) Descriptors based on PCA

Fig. 7.27. Three dimensional scatter plots of descriptors based on Hermite basis functions and PCA.

The last group of morphological descriptors introduced in this section is based on a PCA. The PCA generates a new coordinate system with axes sorted according to the variance of the data. Computing the PCA of a matrix of different QRS complex signals is assumed to return small eingevalues for QRS complexes of normal beats and PACs, while the returned scores of the eigenvectors of the PVCs should significantly differ. In Figure 7.28(a) a superposition of all QRS complex signals of an ECG signal are shown. Figure 7.28(b) illustrates the resulting average QRS complex signal, while Figure 7.28(c) to 7.28(f) present the first four eigenvectors of a corresponding PCA. In addition to that a scatter plot of the scores of the first three eigenvalues is presented in Figure 7.27(b). The group of PVCs is clearly separated from the groups of normal beats and PACs. The scores of the first four eigenvalues are used as morphology depicting descriptors MP_{29} to MP_{32}.

Some more parameters based on the statistical properties of the QRS complexes e.g. skewness or kurtosis are computed. They are described and discussed for the T wave in Section 7.5. Further morphology based descriptors are generated by an analysis of the Wavelet transform of the QRS segments [4]. For the reasons of clarity these descriptors are not introduced here.

7.4.3 Template Generation

The computation of rhythmical and morphological descriptors can be used to classify heart beats into the groups: normal beats, PACs and PVCs. However, one major problem for a classification algorithm is the definition of the boundaries between the groups. Until now, the parameters differ clearly between the different classes of beats within one ECG signal. Yet, the difference between different subjects and different ECG leads was not considered so far. To get reliable classification results using a trained classifier, the proposed descriptors have to be subject independent. All features need to be normalized in the feature space to prevent the classifier from misclassification because of inter subject differences.

Normalisation of Rhythmical Features

In order to detect normal heart beats which can be considered for the normalization of the rhythmical features, the RR intervals in the Poincaré plot are in focus. The RR intervals (Figure 7.29(a)) are transferred by a PCA based method (see Section 3.3) into a new space such that the cloud of normal beats lies in the centre. Thus, the input data P (RR_i and RR_{i+1} in Figure 7.29(a)) is transferred by the PCA into a new space. The resulting scores S are a scaled and rotated version of the input data P (see Figure 7.29(b)). To get the distances Z of each point to the origin of the new

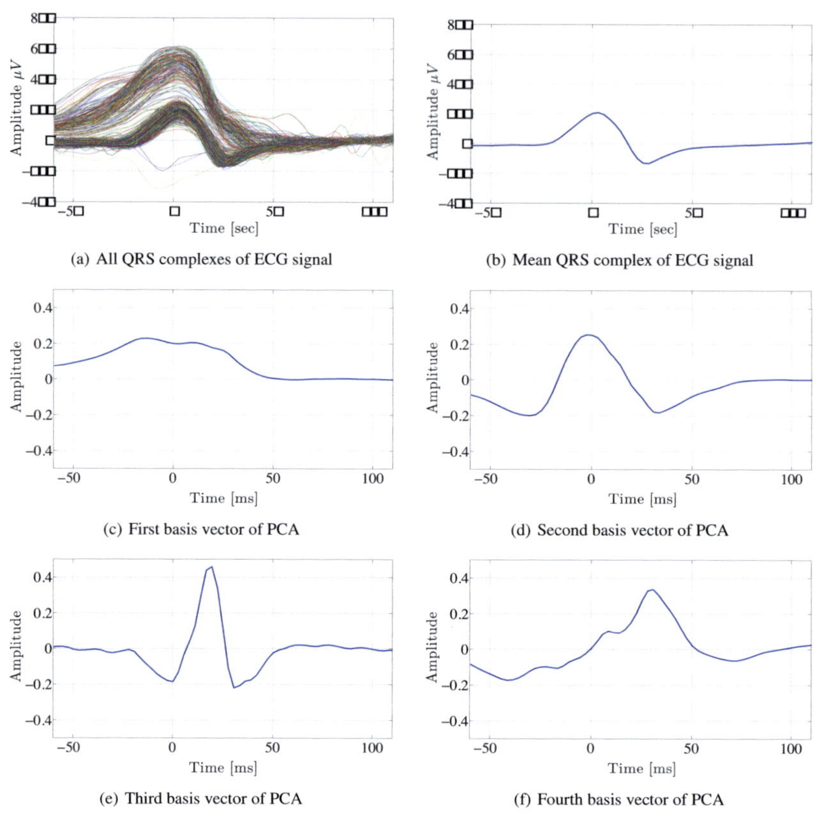

(a) All QRS complexes of ECG signal

(b) Mean QRS complex of ECG signal

(c) First basis vector of PCA

(d) Second basis vector of PCA

(e) Third basis vector of PCA

(f) Fourth basis vector of PCA

Fig. 7.28. PCA based descriptor generation.

space, spanned by PC_1 and PC_2, two distance vectors $\boldsymbol{D}_1$ and $\boldsymbol{D}_2$ are computed by:

$$\boldsymbol{D}_1(i) = Z_{PC1,i}^2 \tag{7.67}$$
$$\boldsymbol{D}_2(i) = Z_{PC2,i}^2 . \tag{7.68}$$

To extract normal heart beats, two thresholds are introduced. The thresholds are based on the variance of the data in $\boldsymbol{D}_1$ and $\boldsymbol{D}_2$. It turned out that the empirical determined thresholds:

$$T_1 = 1.1 \cdot \sqrt{var(\boldsymbol{D}_1)} \tag{7.69}$$

$$T_2 = 1.1 \cdot \sqrt{var(\boldsymbol{D}_2)} \tag{7.70}$$

deliver good results. All scores with values below both thresholds are lying in the 'main' cloud. They all are likely to be normal beats. As the method is iterative, all candidates are used as input data in the next step and the procedure is repeated. The remaining candidates in the 'main' cloud of the 10-*th* iteration are assumed to be normal and used for the normalisation. The resulting 'normal' beats are illustrated in red in Figure 7.29(c).

To normalize the rhythmical features, a mean value $\overline{RP}_k$ and the standard deviation $\widetilde{RP}_k$ is computed for every feature k from the beats previously classified as normal. Equation 7.71 elucidates the normalisation.

$$\boldsymbol{RF}'_k(i) = \frac{\boldsymbol{RF}_k(i) - \overline{RP}_k}{\widetilde{RP}_k}. \tag{7.71}$$

Figure 7.30 shows an example of normalized RR intervals for two ECG recordings of different subjects. It can be seen that the clouds match closely after the normalisation.

Normalisation of Morphological Features

To normalize morphological features, it is obvious to use the differences to the features of a subject specific template. However, a template as the mean of all QRS complexes is not necessary a good representation of the QRS morphology of a normal beat, as the influence of PVCs might significantly change the mean template (see Figure 7.31(a)). Using the beats selected during the normalisation process does not necessarily lead to an exclusion of all PVCs, as PVCs sometimes have normal rhythmical features. A separation process is needed to erase potentially remaining PVCs from the previously selected beats. Figure 7.31(a) shows an example of an ECG recording with PVCs. The mean value of these signals would neither represent normal beats, nor PVCs.

To get a reliable and clear template of the normal beats, the signals are analysed using a PCA. The outlier detection is done based on a Hotelling's T^2 statistic. All beats with a T^2-value in the 80% quantile are considered for the calculation of a

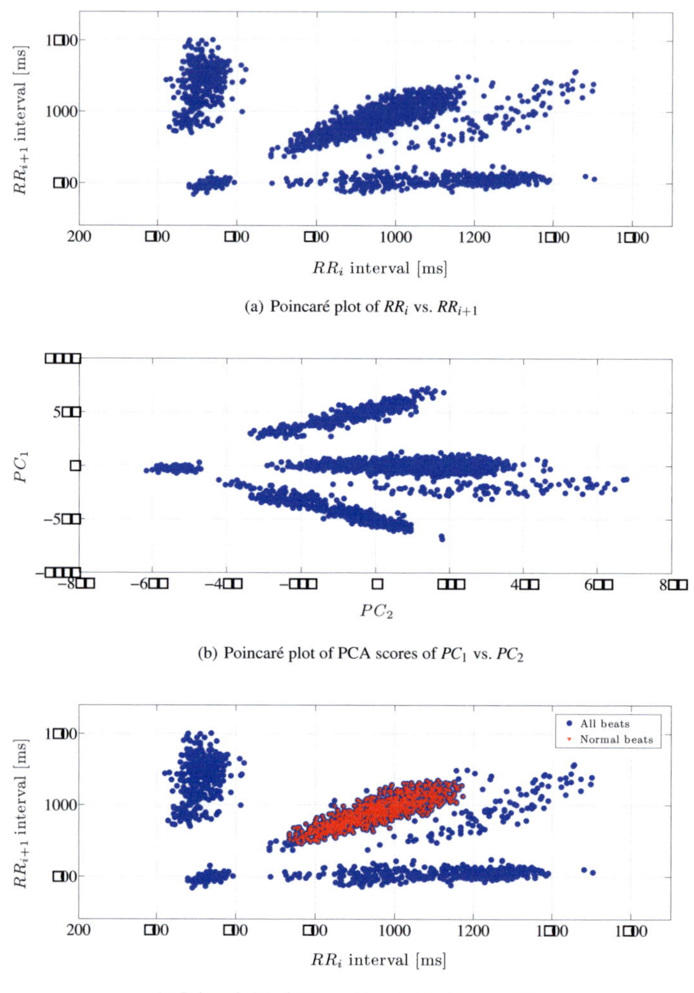

(a) Poincaré plot of RR_i vs. RR_{i+1}

(b) Poincaré plot of PCA scores of PC_1 vs. PC_2

(c) Poincaré plot of RR_i vs. RR_{i+1} and marked normal beats

Fig. 7.29. Illustration of the PCA based detection of normal heart beats.

mean template[5]. Referring to Figure 7.31(a), in Figure 7.31(b) all remaining QRS complexes and the generated template are illustrated.

---

[5] The outlier detection using PCA and Hotelling's T^2 statistics is described in Section 3.3 and additional examples can be found in Section 7.3.1 for the T wave.

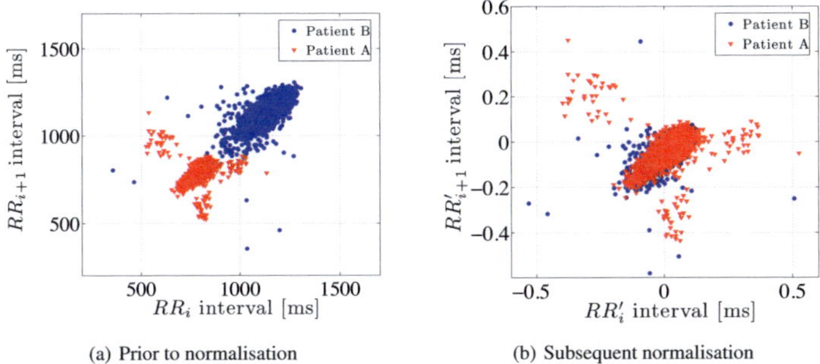

(a) Prior to normalisation

(b) Subsequent normalisation

Fig. 7.30. Poincaré plot of RR_i vs. RR_{i+1} for two different subjects prior and subsequent to normalisation.

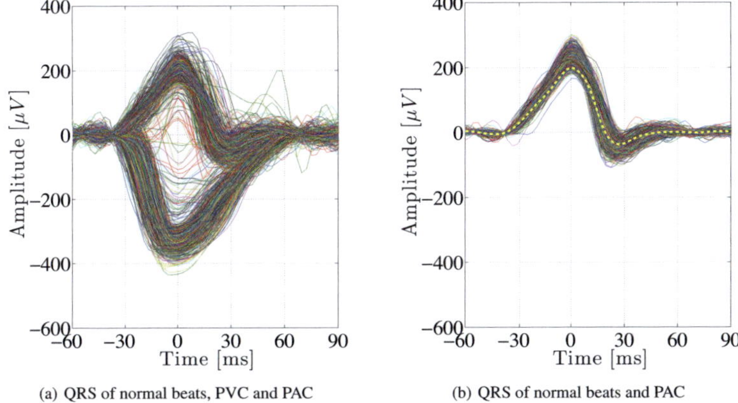

(a) QRS of normal beats, PVC and PAC

(b) QRS of normal beats and PAC

Fig. 7.31. Template generation from QRS complexes of a ECG signal with a huge number of PVCs. a), all detected QRS complexes in the signal. b) remaining QRS complexes used for template generation. Yellow dashed line illustrates the template.

All morphological descriptors are generated for the resulting template (tMP_l). Additionally, out of all beats used to calculate the template, the standard deviation of each morphological feature ($\widetilde{MP_l}$) is calculated. To normalize all morphological features (MP_l) of all beats i, a similar approach is chosen as for the rhythmical features:

$$MP'_l(i) = \frac{MP_l(i) - tMP_l}{\widetilde{MP_k}}. \tag{7.72}$$

Using the normalisation introduced here, it was possible to use the SVM-classifier for different subjects, leads and recording systems without any adaptation.

7.4.4 Classification of the Beats

The classification of the heart beats is done using a Support Vector Machine (SVM) (see Section 3.5). The SVM is a supervised learning method that generates an optimal linear classifier. Similar to many other automatic classification methods, the SVM has to be trained. As the classes which have to be separated by the SVM are not necessarily linearly separable, a kernel function is used to map the input pattern into a new feature space. In this thesis the Gaussian Radial Basis Function (RBF) has been used. For the computation, the algorithm *libsvm* [84], implemented by the Department of Computer Science of the University of Taiwan, was used. In this implementation the kernel function is defined by:

$$K(\boldsymbol{x},\boldsymbol{y}) = \exp(-\gamma \cdot ||\boldsymbol{x}-\boldsymbol{y}||^2), \text{ with: } \sigma^2 = \frac{1}{2\gamma} \tag{7.73}$$

the parameter γ has to be chosen manually. Also the parameter C (for the misclassification error) has to be chosen manually during the training phase.

The range of the parameter values used to classify the heart beats are very different. Larger numeric values might dominate smaller numeric values of other parameters. In addition to that, very large and very small values might cause computational problems. To solve these problems all parameters were rescaled to an interval of -1 to 1. The smallest numeric value was set to -1 while the largest value was scaled to 1.

To find the best values for γ and C, all possible values had to be tested. To optimize the parameter values γ and C, the training process ran in several iterations. In each iteration the values were changed. The quality criteria of the optimization was the Correct Rate (CR) (see Equation 7.74). In the first iterations the number of QRS complexes was small and the span of the parameter values was large. In the following iterations the number of QRS complexes increased. The range for the parameter values of γ and C was set to the ranges with best results in the previous iteration. This was done until the best possible correct rate was achieved.

To train the SVM in this research project the MIT-BIH arrhythmia database, introduced in Section 5.6, was used. The first parameter intervals were set to $\log_{10}(\gamma) \in [-3, -2, -1, \dots 4, 5]$ and $\log_{10}(C) \in [-3, -2, -1, \dots 4, 5]$. The logarith-

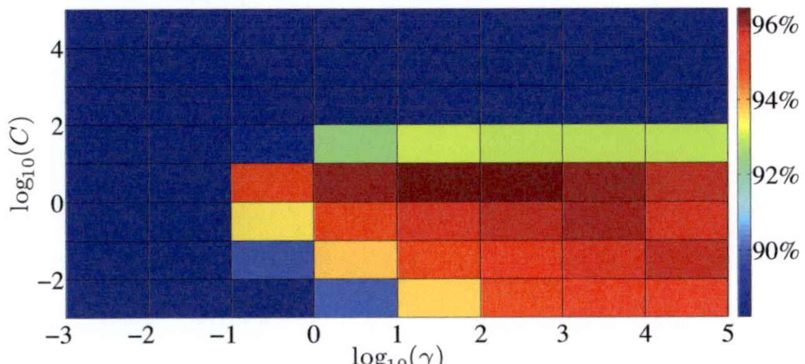

Fig. 7.32. Illustration of the iterative training procedure and adaptation of the free parameters.

mic scaling increase the analysed range. 5000 QRS complexes were considered in the first iteration. Figure 7.32 shows the results of the optimization process. In the next iteration the parameter intervals were chosen to $\log_{10}(\gamma) \in [0;4]$ and $\log_{10}(C) \in [-1;1]$. This was done until 50 % of all QRS complexes of the MIT-BIH arrhythmia database were considered in the training set. Due to the large number of heart beats in the training set, the training phase of the SVM took 14 days.

After the training phase, the SVM was used for classifying ECG data from the MIT-BIH arrhythmia database and from the PVC-study. It turned out that a significant number of normal heart beats were classified as ectopic beats. Visual observations of the misclassified beats showed that noise and artefacts had a high impact on the classification. It became clear that the use of a filter was necessary to further enhance the classification quality (especially for false positive beats). The following three types of filters were tested to denoise the QRS complexes:

- Smoothing filter
- Gaussian filer
- Butterworth filter.

The filters were designed adaptively based on the signal energy in the spectrum. Therefore the spectral distribution of the Fourier transform of the QRS complex is computed. The frequency where 90 % of the total signal energy is included, is set to the cut-off frequency of the filter. Best results have been achieved using the Butterworth filter. Figure 7.33 illustrates the detection of the cut-off frequency.

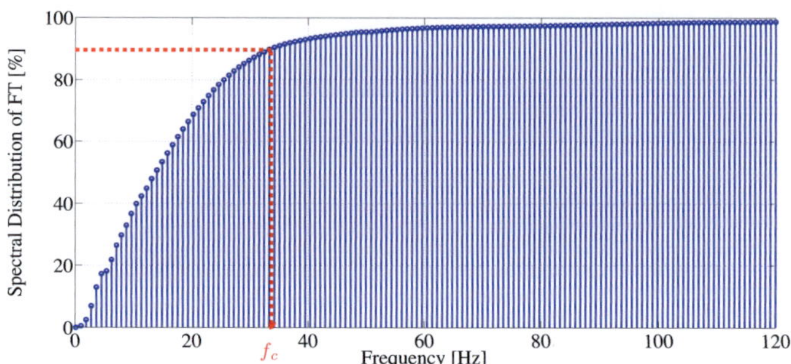

Fig. 7.33. Illustration of the adaptable cut-off frequency of the ECG filter for beat classification.

7.4.5 Results of the Classification

To validate the classification, annotated data from the PVC-study was used. As the SVM was trained on a different dataset (MIT-BIH arrhythmia database) a realistic classification result should be achieved. However, for validation purpose, the following values were calculated:

- Correct Rate:

$$CR = \frac{number\ of\ correctly\ classified\ beats}{total\ number\ of\ beats} \cdot 100\% \qquad (7.74)$$

- Sensitivity:

$$SE = \frac{number\ of\ correctly\ classified\ ectopic\ beats}{total\ number\ of\ ectopic\ beats} \cdot 100\% \qquad (7.75)$$

- Positive Predicted Value:

$$PPV = \frac{number\ of\ correctly\ classified\ ectopic\ beats}{total\ number\ of\ beats\ classified\ as\ ectopic} \cdot 100\% \qquad (7.76)$$

The correct rate is often used for validation, but in a dataset with only few ectopic beats this value is very high if all beats are set to normal. The sensitivity is a good measure for the classification quality of ectopic beats. However, if the number of normal beats classified as ectopic is high, the sensitivity is not affected. The positive predicted value is sensitive forwards this case. Normal beats detected as ectopic lead to a decrease of the PPV.

Besides the validation of the beat classification, the quality of the R peak detection is important. Heart beats which have not been detected during the R peak detection cannot be classified. To prevent the results from a decrease by probably missed or wrong detected beats, only the classification of the correct detected beats were considered for the validation. Additionally the results of the R peak detection are presented. Table 7.2 shows the results for the classification and R peak detection of the whole PVC-study. This study involving 56 subjects and encompasses 290149 heart beats in total. The ECGs have been recorded with eight leads. To improve the classification results, the analysis of all eight channels was additionally tested. All introduced delineation methods in this work are based on one ECG lead. This is necessary, as multi-lead ECG data is not available in all cases[6]. Nevertheless, using the information of all available leads is obvious and advantageous. A multi lead majority decision was used for R peak detection and classification. The results are shown in Table 7.2.

Table 7.2. Results of the QRS detection and beat classification

Channel	Filter	CR_{QRS} [%]	CR SVM [%]	SE SVM [%]	PPV SVM [%]
2	off	99.087	98.314	90.562	44.268
4	off	97.454	98.321	92.164	46.945
6	off	99,752	98.397	90.848	46.873
2	on	99.087	98.797	88.737	51.249
4	on	97.454	98.759	90.930	53.003
6	on	99,752	98.979	92.658	58.972
multi	on	99.864	98.386	86.382	77.245

7.4.6 Discussion

In this chapter a beat classification algorithm based on a SVM was introduced. The classification was based on rhythmical and morphological features extracted from every detected QRS complex. All features were normalized to prevent an adaptation of the classifier due to differences related to heart rate, leads, signal quality or inter subject variability. The SVM was trained using the MIT BIH arrhythmia database. The presented results were achieved by analysing the PVC-study introduced in Section 5.7. For an additional improvement of the classification, a multi-lead analysis has been performed. To measure the quality of the classification,

---

[6] For example the Myocarditis database used in this work has only one ECG lead.

the values of correct rate, sensitivity and positive predicted value are presented. It turned out that the values of *CR* and also for *SE* are satisfactory. The classifier was able to detect most of the ectopic beats correctly. However, the PPV was quite low first. Many normal beats were classified as ectopic. An improvement of falsely classified normal beats could be reached by using an additional signal filter. PPV was increased by about 5% to 12%. However, the best PPV result was still only at 58% and thus not satisfactory. Yet these results could be significantly improved by using a multi-lead decision. PPV was increased to a value of approximately 77%. The value of PPV is still relatively small in comparison to *CR* or *SE*. A comparison to other algorithms could not be presented, as this value is seldom published for a beat classification algorithm. On the other hand, it has to be reasoned that this value only considers ectopic beats. Small numbers of ectopic beats in a signal can lead to a low PPV-value even for a small number of misclassifications. However, the total number of missed ectopic beats is small and to train a new SVM could help to further enhance the results. In fact, the SVM was only trained on the MIT-BIH arrhythmia database. A new SVM trained on data of several different clinical studies is envisioned. Additionally, some modifications in the algorithm, e.g. in the normalisation routines, were recently made. These modifications have not been considered during the training phase. The very time consuming training phase of several weeks made a new training for every change in the algorithm impossible.

To further increase the quality of the classification, the use of additional information based on the P- and the T wave might be considered in the future. The missing of a P wave might be an additional hint for a PAC. The analysis of a subsequent T wave might help to decide whether a normal beat or a PVC was detected. However, at this stage, no P- and T wave information has been included. The T wave delineation algorithm described in Section 7.3 is completely independent from the beat classification. This provides the possibility to use information such as e.g. QT interval length or correlation coefficients of the current heart beat, for the beat classification.

Table 7.2 provides also the results of the validation of the QRS detection algorithm. A reliable detection of the R peak for both, single lead and multi lead detection was achieved. The correct rate of 99,86% for multi lead detection is above average [60].

7.5 Time Series Generation and Wave Extraction for Advanced Signal Processing

ECG delineation, as described in previous sections, is the precondition for the analysis of clinical aspects out of the surface ECG. Points in time, provided by the fiducial point detection, have to be transferred into time series to extract the underlying information. In this work, a framework has been developed to handle time series of the surface ECG. The time series involve either time based information or information on the wave morphology. Additionally, signal parts including whole heart beats or a specific part of the heart beat, e.g. the T wave can be extracted for every beat of the ECG signal.

7.5.1 Time Series Generation

The generation of time series out of the ECG data is an important step for the analysis of ECG data. Time series involve information extracted out of every single heart beat which are written into a vector. In every element of the vector, the descriptor value of the corresponding heart beat is stored. Most common time series out of the ECG is the heart rate, e.g. in terms of a series of RR intervals.

In this work, time series out of the ECG are spread into two groups:

- ***Time based:*** Data out of time based descriptors like e.g. the RR- or the QT interval.
- ***Morphology based:*** Data out of descriptors describing the morphology of a wave of the ECG.

7.5.1.1 Time Based Descriptors

Time based descriptors used in this work are:

RR RR intervals, in milliseconds
QT QT intervals, in milliseconds
RT RT intervals, in milliseconds
QT_{cB} QT intervals, corrected with formula of Bazett
QT_{cF} QT intervals, corrected with formula of Fridericia
δ_{I-III} Time intervals of the T wave

The RR interval of heart beat i is calculated by:

$$RR(i) = R_{i+1} - R_i \tag{7.77}$$

The QT interval of heart beat i in this work is given by[7]:

$$QT(i) = T_{end,i} - Q_{peak,i} \tag{7.78}$$

Similar to the QT interval, the RT interval (RT) can be defined as:

$$RT(i) = T_{end,i} - R_i \tag{7.79}$$

As described in Chapter 8.2.1, the QT interval has to be heart rate corrected. Two standard correction formulas (Bazett and Fridericia) have been used to generate time series:

$$QT_{cB}(i) = \frac{QT(i)}{\sqrt{\frac{1}{1000} \cdot RR(i)}} \tag{7.80}$$

$$QT_{cF}(i) = \frac{QT(i)}{\sqrt[3]{\frac{1}{1000} \cdot RR(i)}} \tag{7.81}$$

Both, formula of Bazett [80], Equation 7.80, and formula of Fridericia [86], Equation 7.81, need to have RR interval values in seconds. As all time based values are computed with the unit milliseconds, the RR value has to be multiplied by $\frac{1}{1000}$.

---

[7] In some works the QT interval i is related to the RR interval $i-1$, [85]

Used T wave time intervals δ_I to δ_{III} can be seen in Figure 7.34. The computation is done by:

$$\delta_I(i) = T_{end,i} - T_{peak,i} \tag{7.82}$$

$$\delta_{II}(i) = T_{end,i} - T_{onset,i} \tag{7.83}$$

$$\delta_{III}(i) = T_{\xi_2,i} - T_{\xi_1,i} \qquad \text{with the inflexion point } T_\xi \tag{7.84}$$

All descriptor values are computed for every single heart beat in the ECG signal.

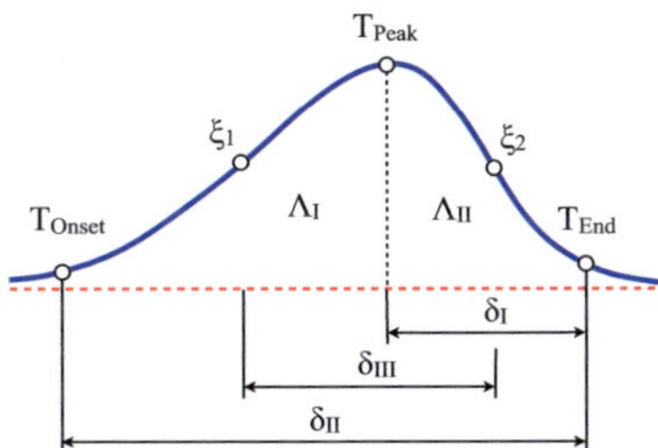

Fig. 7.34. Illustration of time and area based T wave descriptors. The left half of the T wave is represented by Λ_I, the right half of the T wave is represented by Λ_{II}, while the sum of both is represented by Λ_{III}. The baseline is illustrated by the red dashed line. Time intervals are shown as arrows under the plot.

7.5.1.2 Morphological Based Descriptors

Morphological descriptors used in this work:

Ψ	Normalisation factor for the T wave
μ'	Expected value of the T wave
σ'	Standard deviation of the T wave
γ'_1	Skewness of the T wave
β'_2	Kurtosis of the T wave
$\alpha'_{I/II}$	Gradients at the inflexion points of the T wave
v	Difference between baseline and maximum of the T wave
Λ_{I-III}	Area under the T wave (time integral)

These morphological descriptors have been developed for the T wave only. In many standard leads of the surface ECG, the T wave has a shape similar to a Gaussian function. Gaussian functions are well known in probability theory and statistics. A description by expected value, standard deviation and central moments is common. Thus, a description of the T wave using the equations of statistics is self-evident.

The T wave values, $\mathbf{T}(t)$ are interpreted as a density function for all following descriptors. The names of those descriptors are signed by an additional dash. A normalisation of the vector $\mathbf{T}$ has to be done to compute the statistical values. The normalisation is mathematically described in Equation 7.86. In this equation $\mathbf{T}(t)$ is time dependent. The values of t are the time stamps starting from t_1 and ending at t_e.

$$\Psi \cdot \sum_{t=t_1}^{t_e} \mathbf{T}(t) \stackrel{!}{=} 1 \tag{7.85}$$

$$\mathbf{T}' = \Psi \cdot \mathbf{T}(t) \tag{7.86}$$

The normalisation factor is unitless. Out of the normalized T wave $\mathbf{T}'$, an expected value can be calculated by:

$$\mu' = \sum_{t=t_1}^{t_e} t \cdot \mathbf{T}'(t) \tag{7.87}$$

The expected value is a measure of the centroid of the T wave. The value is time based, thus unit is milliseconds. Out of the expected value, the standard deviation (σ') is calculated:

$$\sigma' = \sqrt{\sum_{t=t_1}^{t_e} (t - \mu')^2 \cdot \mathbf{T}'(t)} \tag{7.88}$$

σ' is a measure for the variance of the T wave samples around the expected value. The unit is again time based, (milliseconds).

The third central moment, the skewness (γ_1') is mathematically described for the T wave in Equation 7.89

$$\gamma_1' = \frac{\sum_{t=t_1}^{t_e} (t - \mu')^3 \cdot \mathbf{T}'(t)}{\sigma'^3} \tag{7.89}$$

Skewness is a measure of the asymmetry of the wave. The value of γ_1' lies around zero. This descriptor is unitless. The last descriptor out of the statistics is the fourth central moment, kurtosis (β_2).

$$\beta_2' = \frac{\sum\limits_{t=t_1}^{t_e} (t-\mu')^4 \cdot \mathbf{T}'(t)}{\sigma'^4} \tag{7.90}$$

Kurtosis is a measure of the 'peakedness' of a function. The descriptor is unitless. Its values are spread around three.

Two further descriptors α_1' and α_2' showing the gradient at the inflexion points ξ_1 and ξ_2 of the first and second half of the normalized T wave are introduced. These parameters do not come from the statistics, but as they have been calculated on the normalized wave they have been also marked by an additional dash.

$$\alpha_I' = \frac{\mathbf{T}'(\xi_1) - \mathbf{T}'(\xi_1 - 1)}{t(\xi_1) - t(\xi_1 - 1)} \tag{7.91}$$

$$\alpha_{II}' = \frac{\mathbf{T}'(\xi_2) - \mathbf{T}'(\xi_2 - 1)}{t(\xi_2) - t(\xi_2 - 1)} \tag{7.92}$$

The following descriptors are extracted from the original (not normalized) waves. A standard descriptor is the height of the wave v. The definition is quite simple, the difference between the baseline and the maximum of the wave.

The last analysed morphological descriptors are values representing an area under the curve. Three different curve intervals have been generated to calculate areas under the curve. Figure 7.34 illustrates the three areas. The descriptors are named as Λ_{I-III}. The computation is described in Equation 7.93. The corresponding boundary values t_a and t_b can be found in Figure 7.34.

$$\Lambda = \Delta t \cdot \sum\limits_{t=t_a}^{t_b-1} \frac{\mathbf{T}(t) + \mathbf{T}(t+1)}{2} \qquad \Delta t = \frac{1}{f_s} \tag{7.93}$$

As the descriptors are computed for every single heart beat and T wave respectively, a vector can be generated out of the descriptor values. Every vector can be seen as a time series.

All generated time series are not equidistant in time. The duration between two heart cycles, changes for every heart beat and thus the time distance between each value is varying. This can be important in a further signal processing, as it was done in Chapter 9. Hence, a vector t_{TS} is defined involving the time stamps of the time series values. The time stamp is equal for all descriptors. Date of the corresponding R peak in the ECG signal is used.

$$t_{TS}(i) = R_i \tag{7.94}$$

The time stamps are extracted in milliseconds.

7.5.2 ECG Wave Extraction

To analyse the morphology of ECG cycles, whole heart beats, QRS complexes and T waves can be extracted. The waves are cut out of the ECG signal and are written into a matrix Θ. All waves have to be aligned against each other.

7.5.2.1 Whole Heart Beat

For whole heart beats the signals are aligned according to their R peak. Figure 7.16 shows a visualisation of a matrix involving whole heart beats of an ECG. In every line j of the matrix Θ_{HB}, the amplitude values of the ECG signal of one heart beat is listed. The signal parts are all of equal length.

$$\Theta_{HB}(j) = [ECG(R_i - n_1), ECG(R_i - n_1 + 1), \ldots \\ , ECG(R_i + n_2 - 1), ECG(R_i + n_2)] \tag{7.95}$$

Where n_1 is the time before R_i and n_2 is the time distance behind R_i. Figure 7.16 shows a matrix with exported heart beats.

7.5.2.2 QRS Complex

The QRS complex can be extracted to matrix Θ_{QRS} by shortening the length of n_1 and n_2 in Equation 7.95. Figure 7.35 shows a visualisation of the extracted QRS complexes of a Holter ECG signal.

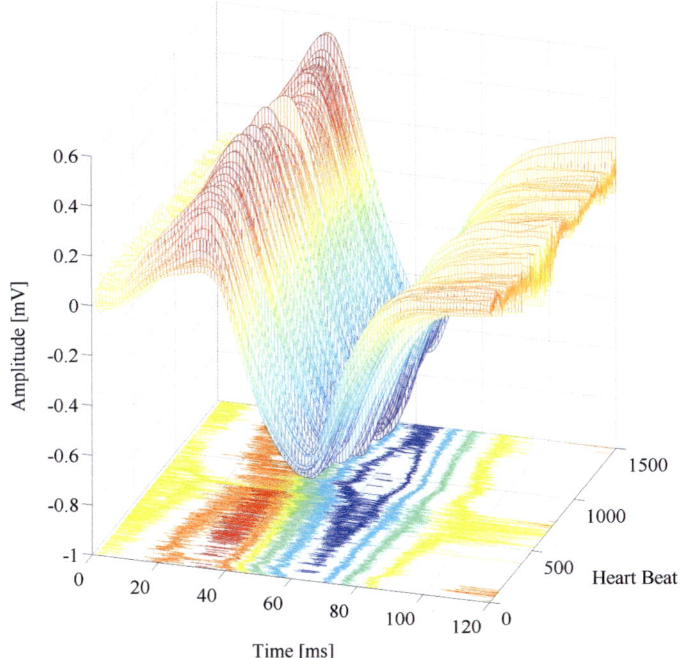

Fig. 7.35. Visualisation of QRS complexes in matrix Θ_{QRS}

7.5.2.3 T wave

The T wave matrix Θ_T has to be generated by aligning the T waves at T_{peak}. This is not as easy as in case of the QRS complexes. The peak of the T wave is smoother than the R peak. Noise on the wave might lead to a wrong alignment, if exactly the highest point is used. To prevent the waves from small dislocations, the maximum of the smoothed waves has been used. A further improvement was reached by a correlation alignment algorithm.

Refinement of T wave alignment

A mean T wave T_0, out of the smoothed and aligned T waves Θ_T is generated. This mean wave is then cross correlated with every single wave in a small region around T_{peak}. The waves are aligned to the position with the highest cross correlation coefficient.

After the refinement, a matrix $\widehat{\boldsymbol{\Theta}}_T$ results with best possible alignment of the waves. Figure 7.36 illustrates a matrix of aligned T waves.

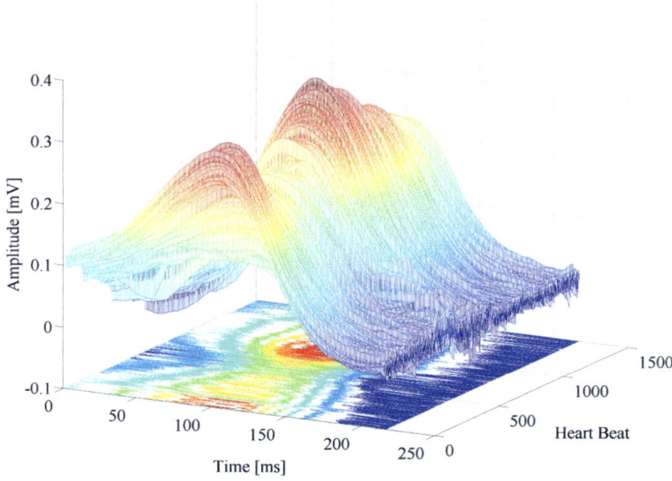

Fig. 7.36. Visualisation of T waves in matrix $\widehat{\boldsymbol{\Theta}}_T$

Outlier Detection

Holter ECG signals are often disturbed by noise and artefacts. The T wave is prone for significant morphology changes due to noise and artefacts. Hence, it is necessary to detect outlier waves.

In Section 7.3.1 an outlier detection routine has already been described. This outlier detection was designed with very hard thresholds, as in the context of template generation, even small outliers have to be detected. The relative high number of T waves and the fact of computing a template as a mean of the remaining T waves, made this a good choice. In the particular case, the number of outliers has to be as small as possible. Only waves touched by noise and artefacts have to be thrown out. Waves influenced by changes of the heart rate or any physiological reasons need to remain.

The outlier detection is made in two steps. First, a mean wave $\hat{\boldsymbol{T}}_0$ out of $\widehat{\boldsymbol{\Theta}}_T$ is generated. For every T wave in matrix $\widehat{\boldsymbol{\Theta}}_T$ a Pearson correlation coefficient according to Equation 7.28 is computed. All waves having a correlation coefficient smaller than 0.9 are classified as outliers. Mainly T waves disturbed by artefacts should be detected. A new matrix $\widehat{\boldsymbol{\Theta}}'_T$ results.

Next, T waves influenced by noise and signal jitters are in focus. To detect those waves, the mean wave of $\widehat{\boldsymbol{\Theta}}'_T$, $\hat{\boldsymbol{T}}'_0$ is subtracted from all waves. The sum of all absolute values of the signal minus $\hat{\boldsymbol{T}}'_0$ is calculated. To be independent of the sampling frequency, the result is divided by the number of sample points N. Equation 7.96 shows the computation for T wave j of the matrix.

$$k(j) = \frac{1}{N} \sum_{n=1}^{N} \left| \boldsymbol{T}_j(n) - \hat{\boldsymbol{T}}'_0(n) \right| \tag{7.96}$$

All T waves having a k-value higher than fife times the mean of all k-values, are also classified as outliers and withdrawn from $\widehat{\boldsymbol{\Theta}}'_T$. A matrix $\widehat{\boldsymbol{\Theta}}''_T$ results finally.

Interval Analysis

Until now, all values and waves are considered for every single heart beat. Hence, even beat to beat changes in the ECG data are observable.

To analyse slow variations in the ECG, as they can be caused by a medication, analysing mean values out of time intervals is more beneficial than a beat to beat observation. Additionally, using intervals the influence of short episodes of low quality ECG can be reduced. Thus, a more reliable analysis is possible.

In this work, intervals have been generated for all time based parameters of the ECG and the descriptors based on the T wave morphology. In case of RR, QT and QT_c, simply the mean of all parameter values inside the interval was calculated.

An example for the I-*th* interval of the time series $\boldsymbol{RR}$ is presented in Equation 7.97.

$$\overline{\boldsymbol{RR}}(I) = \frac{1}{l_{2,I} - l_{1,I}} \cdot \sum_{j=l_{1,I}}^{l_{2,I}} RR(j) \tag{7.97}$$

In Equation 7.97, $l_{1,I}$ represent the number of the first and $l_{2,I}$ the number of the last wave of the interval I.

In case of the morphology based descriptors, a template as the mean of all T waves inside the interval is generated. The corresponding parameter values are calculated using this template. Computation of the template's parameters is equal to the computation of the parameters of a single T wave. Equation 7.98 shows the computation of the template for interval I.

$$\overline{T}_I = \frac{1}{l_{2,I} - l_{1,I}} \cdot \sum_{j=l_{1,I}}^{l_{2,I}} \widehat{\boldsymbol{\Theta}}_T''(j) \tag{7.98}$$

$$\tag{7.99}$$

Time series and matrices, representing the mean values of all intervals are labelled with a bar on top to distinguish them to time series and matrices of single beat values. The matrix involving the mean waves of all intervals is named as $\overline{\boldsymbol{\Theta}}_T$.

7.5.3 Export Structure

Analysing Holter ECG data in clinical studies leads to a huge amount of different data. Every patient in the study can have some personal information such as age, sex, or findings. Several time series can be generated from the ECG data after fiducial point detection as described in Section 7.5.1. Beside the one dimensional time series vectors, matrices with ECG signal segments such as T waves or QRS complexes, can be extracted.

To handle these different types of data, a Matlab structure has been developed which contains all signals and information. This *export structure* called file has equal structure for all participants of an ECG study. Thus, automatic processing of complete study data is possible.

Figure 7.37 shows a schematic of the export structure. In the last column an example or description of the data is given. The structure cannot be modified, but entries, not available for a patient, must not exist. For automatic analysis of several subjects, only the existence of the data which is in focus has to be proofed by an algorithm. Values shown in Figure 7.37 are the basic values which can be found in every export structure. Additional patient related information stored in a *unisens.xml* file of a record are automatically added.

For all parameters and waves, a time vector can be found in the structure to get the corresponding time in the ECG signal. Time stamps are set to the time of R peak (R_i) of the corresponding heart beat.

All time series are generated during the export and not during fiducial point detection.

Layer I	Layer II	Layer III	Layer IV	Layer V	Values
Record	Information Table				*String Data (Cell, N x 5)*
	Fiducial Point Table	Data			*Matrix M x 16*
		Colom String			*String Data (Cell, 16 x 1)*
		Beat Type String			*String Data (Cell, 2 x 20)*
	Name				*Tobias Baas 01*
	Export Date				*01-Dec-1980 07:26:00*
Patient	Study				*IBT 01*
	Clinical Research No				*C001*
	Date of Record				*01-Dec-1980*
	Starttime				*08:00*
	Patient Name				*Tobias Baas*
	Patient Age				*31*
	Patient Sex				*Male*
	Analyte				*none*
	Findings				*none*
	Others				*none*
ECG	Time Series (ts)	RR	Time		*Vector (1 x M)*
			RR Data		*Vector (1 x M)*
			HR Data		*Vector (1 x M)*
		RT	Time		*Vector (1 x M)*
			Data		*Vector (1 x M)*
		QT	Time		*Vector (1 x M)*
			Data		*Vector (1 x M)*
			RR Data		*Vector (1 x M)*
			HR Data		*Vector (1 x M)*
			QTc Bazett Data		*Vector (1 x M)*
			QTc Fridericia Data		*Vector (1 x M)*
	Waves	QRS Complex	Time		*Vector (1 x M)*
			Data		*Vector (L x M)*
			Signal Information		*String Data (Cell, 13 x 1)*
			Lead Number		*Number*
			Lead Name		*String*
			File Name		*String*
		Heart Beat (HB)	Time		*Vector (1 x M)*
			Data		*Vector (L x M)*
			Signal Information		*String Data (Cell, 13 x 1)*
			Lead Number		*Number*
			Lead Name		*String*
			File Name		*String*
		T wave	Time		*Vector (1 x M)*
			Data		*Vector (L x M)*
			Data Normalized		*Vector (1 x M)*
			Signal Information		*String Data (Cell, 13 x 1)*
			Lead Number		*Number*
			Lead Name		*String*
			File Name		*String*
			Different Morph. Descritors		*Each, Vector (1 x M)*
			Intervals	Start Time of Interval	*Vector (1 x I)*
				Data	*Vector (L x I)*
				Data Normalized	*Vector (1 x I)*
				Different Morph. Descritors	*Each, Vector (1 x I)*

Fig. 7.37. Schematic of the export structure of BSAT

7.6 BioSignal Analysis Toolbox BSAT

BSAT is a software toolbox which was developed during this research work. The software helps to analyse ECG and blood pressure data. It provides a number of classical analysis tools, such as adjustable signal filter or ECG delineation algorithms. The software is developed for evaluation of ECG and Blood pressure data. It provides an information management system to store clinical relevant information at any position in the data. The graphical user interface provides a fast and detailed signal display, which allows manual analysis of any signal part regardless of sample rate or resolution.

Figure 7.38 shows a eight lead Holter ECG in BSAT. The automatically detected fiducial points can be seen as vertical lines. The software was developed to view

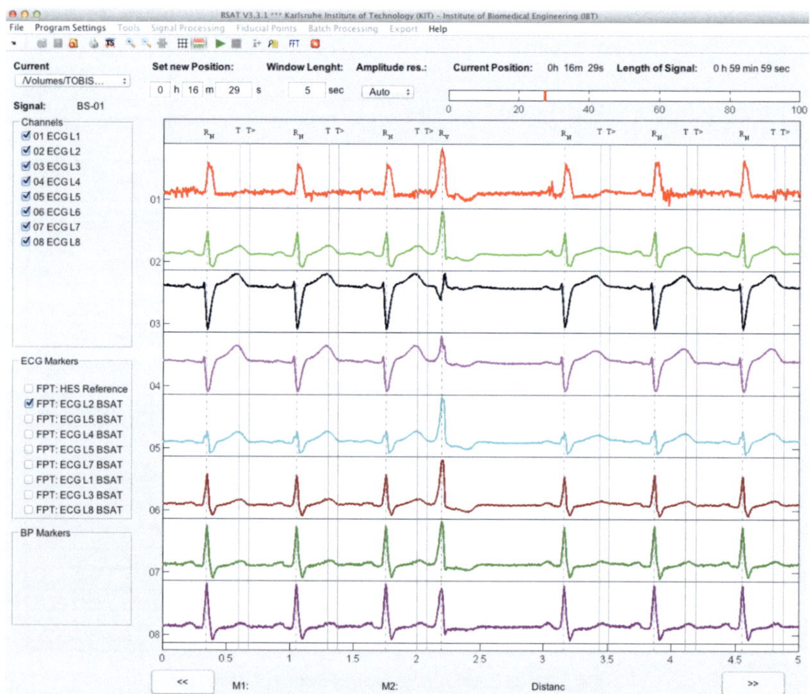

Fig. 7.38. Screenshot of the Main-GUI of the software toolbox BSAT.

and visually analyse ECG and blood pressure data. Any part of a signal can be shown by a simple mouse click, regardless of resolution, number of leads or sample rate. Information of the patient and recording are easily available. New information and supplementary notes can to be stored, viewed and changed by any user of the program.

To evaluate clinical studies, the software has a batch processing routine. The opportunity to save settings which are reloaded during the batch processing makes the work on complete studies possible. An export of information, signals and results is implemented for advanced signal analysis.

The software was used for all investigations in this research work.

Main features of BSAT summarised:

- Denoising
- Fiducial point detection
- Beat classification
- Time series generation
- Data and signal export
- Batch processing
- Information system
- Unisens 2.0 compatible

Data Handling

To open files in BSAT a signal in Unisens 2.0 [49] format is needed. Unisens 2.0 file format is able to handle a various amount of signal data in one folder. This makes the data handling easy. Additional personal information on the patient, the recording system, dates and times can be stored in the xml file of the Unisens file folder.

New data can be read into BSAT by generating a Unisens 2.0 folder. To do so, either the data available as .mat file or raw ECG data out of the recording system *BSAT-1012* are required. The import can be done using the import assistant in the BSAT software.

Filtermask

The filter described in 7.1 can be set up over a filter mask in BSAT. Figure 7.39 shows the GUI to setup the filter. A storage system is available to save the filter setup. During a batch processing the stored setup can be reloaded.

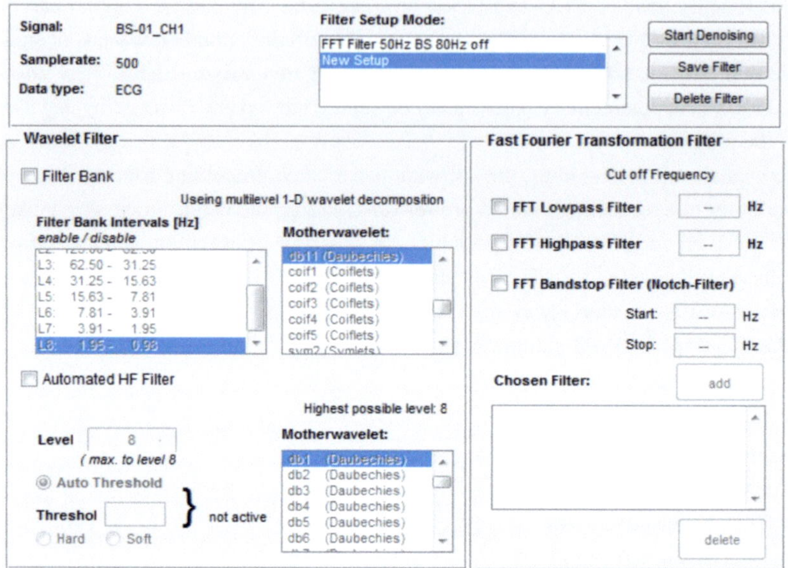

Fig. 7.39. Screenshot of the filter settings mask of BSAT. The filter mask is classified into Fourier and Wavelet based filters.

Information Management System

In the information management system, information related to the recordings can be stored. These measurement related information is always based on a time point in the signal. All information related to a signal point can be stored in the notes. An example are the measurement results of drug blood plasma concentrations measured at some points during an ECG measurement. The times of eating and drinking can be registered. Comments from a physician which might be interesting for an engineer could also be stored in the information notes. Hence an easy way of information exchange is provided by the information management system of BSAT.

Figure 7.40 shows information notes of BSAT with listed events of PVC. A visualisation of a PVC event in the ECG signal is available.

Batch Processing

The processing of complete datasets is necessary to evaluate clinical data. BSAT provides batch processing for denoising, fiducial point detection, beat classifica-

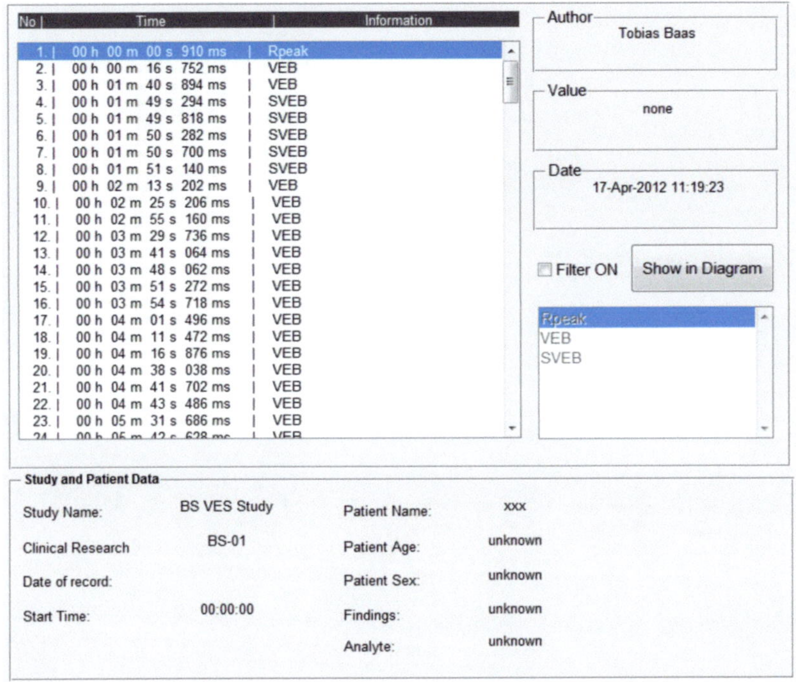

Fig. 7.40. Screenshot of the information management system of BSAT. The list involves detected PVCs. Below the list, additional information can be seen. The entries of the list can be shown in the signals diagram.

tion and the export of signals and time series.

During the batch processing a stored setting is used and all recordings are processed in the same manner.

Part III

Investigations

8

Drug Induced Repolarisation Changes

8.1 Introduction

In the last two decades, several drugs have been withdrawn from the market as reports of sudden cardiac death associated with the drug have been published [87, 88]. The genesis of antiarrhythmia resulting into a potentially fatal ventricular tachyarrhythmia, known as Torsade de Points (TdP), is associated with the prolongation of the QT interval [89]. Hence, the QT interval, representing the duration of the systolic interval, is in focus as a potential risk factor for humans in drug safety studies [88]. Not only antiarrhythmic drugs have been associated with TdP, also non cardiac drugs arouse suspicion to be responsible for TdP. Some few reports on cardiotoxicity of anti-psychotics and first generation histamine H_1-receptor antagonists in the 1960s and 1970s can be found. In this time, the ability of non cardiac drugs to prolong the QT interval was seen as a pharmacological 'curiosity' of incalculable clinical relevance [87].

Until today, the relationship between QT prolongation and proarrhythmic effects of drugs is not fully understood. The prolongation of the QT interval is often observed in clinical studies, while the observation of TdP is very seldom. This might be caused by the relative small number of participants in clinical trial studies I to III [87].

Preclinical studies can give an indication for 'proarrhythmical danger'. However, the information out of preclinical studies is only poorly understood, which might be reasoned by the non uniform methodology of those studies [90].

Until now, most important risk factor for drug safety studies is the heart rate corrected QT interval. Thus, pharmaceutical companies and the Food and Drug Administration (FDA) responsible for drug approval, are focused on the QT interval evaluation of clinical trial studies.

On the International Conference for Harmonization (ICH), E14 guidance [91], which recommends a thorough QT (tQT) study in clinical development to detect an arrhythmogenic potential, was introduced. The focus of those studies is on clinically relevant drug induced QT prolongation. The analysis of the difference between the QT intervals of a subject under the influence of a drug, relative to placebo can be one example. [92]

In the E14 guideline, hard thresholds for evaluating thorough QT/QTc studies are presented. The result of a thorough QT study gives direction on the amount of information to be collected in the following development steps of the compound. Subsequent definitions are made in [91]:

- A negative *'thorough QT/QTc study'* will almost always allow the collection of on-therapy ECGs in accordance with the current practices in each therapeutic area to constitute sufficient evaluation during subsequent stages of drug development.

- A positive *'thorough QT/QTc study'* will almost always call for an expanded ECG safety evaluation during later stages of drug development.

The regulations regarding to a prolonged QTc interval, have one main criterion: 'Drugs that prolong the mean QT/QTc interval by around five milliseconds or less do not appear to cause TdP. ... A negative thorough QT/QTc study is one in which the upper bound of the 95% one-sided confidence interval for the largest time-matched mean effect of the drug on the QTc interval excludes 10 ms' [91].

Only a short section is focused on morphology analysis in the FDA regulatory guideline. There are no special advices, requested information or any work flows, which have to be considered yet. However, the analysis of morphological features of ECG waves is desirable, since the QT interval alone is not a reliable marker for drug induced arrhythmias [93].

8.2 Drug Induced QT Prolongation

The focus of this section is on drug induced prolongation of the heart rate corrected QT interval. The prolongation of QT_c due to Moxifloxacin and an unknown compound (UnC) are analysed on two studies from THEW, named tQT_I- and tQT_{II}-study.

8.2.1 QT Correction

To analyse the effects of drug induced QT prolongation, the heart rate dependent QT interval has to be corrected. Most common heart rate correction formula in clinical routine is the formula of Bazett (Equation: 7.80) [89, 90]. Bazett divided the QT interval of a heart cycle by the square root of the RR interval [80]. The formula of Bazett standardizes the measured QT interval to a heart rate of 60 bpm. The formula of Bazett is highly controversial, as it has been identified to artificially prolong the corrected QT interval QT_c for heart rates above 60 bpm and shorten QT_c for heart rates below 60 bpm [94]. Hence, a various number of QT correction formulas has been developed. They can be divided into two groups:

- The first group, to which Bazett's formula belongs, uses the current heart cycle to calculate a corrected QT_c. A time lag between the adaptation of the QT interval to an increased or decreased RR interval (QT/RR hysteresis) [95], is not considered by those correction formulas. Tabo in [96] compares several common correction formulas and analyses their ability to show the effect of drug induced QT prolongation in dogs.
- The second group considers the QT/RR hysteresis and its time lag. Several different approaches have been made for dynamic modelling of the QT/RR coupling. To analyse the RR- and QT interval, windowing is very common. Pueyo et al. in [97] compare several window-based correction formulas. In this and other works, the QT interval of the ith heart beat is related to a history of previous weighted RR intervals.

Other approaches are based on a Transfer Function (TRF) formalism to describe the static and dynamic coupling of the QT/RR relationship [98].
In 2008, several dynamic models based on heart cell measurements and simulations have been introduced by the author [99]. To develop the models, heart cell measurements have been made at Pfizer Research and Development, Sandwich UK. The same stimulation protocols have been used to simulate action potentials of heart cells at the Institute of Biomedical Engineering (IBT) Karlsruhe, Germany. The relationship of Action Potential Duration (APD) and Basic Cycle Length (BCL) was used to develop the models. Some of the introduced models used Bazett's formula as a static relationship and added a dynamic relationship on top [100]. A detailed report on the work can be found in [101].

The introduced publications in the field of QT/RR coupling are just a small sample. Various other models and correction formulas can be found in literature. However, a golden standard has not been found yet. Every approach has disadvantages and none of them can be used in every clinical field. Moreover, QT/RR relationship seems to be highly individual even among healthy subjects. Malik in [102] draws the conclusion: 'For detailed precise studies of the QT_c interval (for example, drug induced QT interval prolongation), the individual QT/RR relation has to be taken into account.'.

To evaluate thorough QT studies of the telemetric and Holter ECG Warehouse (THEW), subject specific QT/RR relation was not available. Finding a QT correction method satisfying all points described above was not part of this work. The author decided to use the formula of Bazett with its long tradition in the clinical trial evaluation.

8.2.2 Methods

Holter ECG data of both THEW studies, tQT_I and tQT_{II} has been delineated by the methods introduced in Section 7.2. The delineated ECG data was used to generate time series of QT and RR intervals as described in Section 7.5. The data of every ECG recording was divided into intervals of 30 minutes for tQT_I-study and intervals of one hour for tQT_{II}-study. For each time interval and recording a, representative corrected QT interval was calculated as the mean of all corrected QT intervals, as described in Section 7.5.2.3, Equation 7.97.

Measurement data of every participant exists of two different measurements with equal "environmental" parameter like for example length of recording or daytime of the beginning. In one measurement a drug and in the other a placebo was administered at a fix time after the recording was started. The sequential arrangement of drug- and placebo recording was random and unknown for both, participants and physicians.

To prevent the influence from any environmental factors, the measurement process has been standardised. Factors like starting time, lunch- and dinner time have been kept fixed for all measurements. To detect the influence of the drug, a comparison of corresponding time intervals of both recordings, aligned with dose time[1] is reasonable.

---

[1] In this work, time of drug or placebo administration is named as *"dose time"* or *"time of dose"*. Moreover, *"pre dose"* represents the time before drug administration and *"post dose"* accordingly the time after drug adminsitration.

To prevent inter-day variability, the values resulting out of the last interval pre dose have been subtracted from all intervals of the measurements. Since two differences are calculated, once between the time intervals of both measurement days and once between the dose interval and all other intervals, the results are labelled as $\Delta\Delta$ (double delta). The calculation is explained in Section 8.2.3 for a pseudo descriptor χ.

8.2.3 Calculation of $\Delta\Delta$-values

At this point, the workflow of calculating $\Delta\Delta$-values is shortly explained. The mathematical formulation is valid for all descriptors. Equation 8.1 and 8.2 show the calculation for a pseudo descriptor χ for Δ and $\Delta\Delta$ analysis, as they have been used in this work. The values of an interval k for one subject are introduced. $\overline{\chi}_{Mox}$ refers to the mean values from a Moxifloxacin recording, accordingly $\overline{\chi}_{Plo}$ represents the corresponding mean values of a placebo measurement. The calculation of the mean values was shown in Section 7.5.2.3. The mean value of the last pre dose interval is labelled as χ_{Mox-PD} and χ_{Plo-PD}, respectively.

$$\Delta\chi(k) = \overline{\chi}_{Mox}(k) - \overline{\chi}_{Plo}(k) \tag{8.1}$$

$$\Delta\Delta\chi(k) = (\overline{\chi}_{Mox}(k) - \chi_{Mox-PD}) - (\overline{\chi}_{Plo}(k) - \chi_{Plo-PD}) \tag{8.2}$$

All $\Delta\Delta$-values returned by Equation 8.2 are grouped for corresponding intervals of all subjects. Hence, the boxplots represent the distribution of the $\Delta\Delta$-values of all subjects of the study in one time interval.

Boxplots

Figure 8.1 illustrates the information readable from the boxplots presented in this work. Outliers have not been shown in the boxplots for clearness reasons.
Triangles in the boxplots mark compare intervals. Compare intervals can be used as a visual significance test. If the compare intervals between the two triangles in a box do not overlap within two groups, it can be assumed that both groups have different medians by a significance level of at least 5 %.

8.2.4 Results

To illustrate the prolongation of the QT_c interval, boxplots are shown. In every time interval, the QT_c-values of every participant of the study is combined and

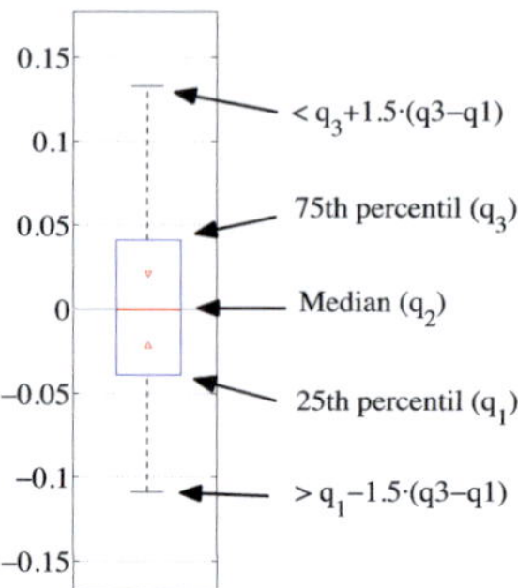

Fig. 8.1. Information indicated in a boxplot

statistically evaluated. These statistics can be seen in the boxplots. The formula of Bazett [80] was used for heart rate correction.

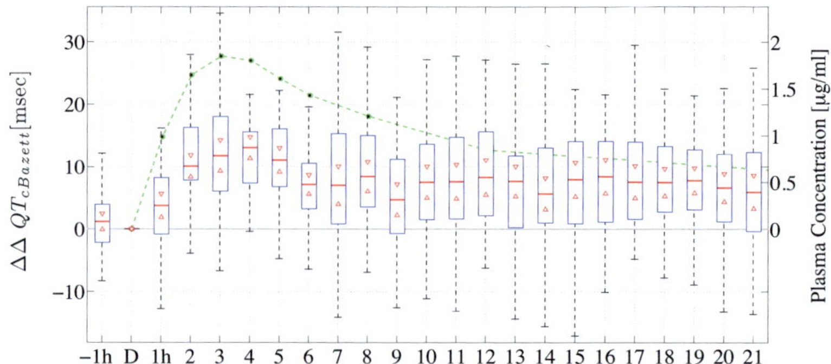

Fig. 8.2. QT prolongation due to Moxifloxacin in the clinical trial tQT_{II}-study. The boxplot includes the data of 57 evaluated participants of the study. A significant increase in the corrected QT interval of about 12 ms can be observed after the administration of Moxifloxacin. The blood plasma concentration of the medication is shown in green.

At first, the focus is on the THEW tQT_{II}-study. These measurements involve information on the blood plasma concentration of the drug Moxifloxacin. 57 subjects of the study have been taken into account. Some more information on both THEW studies are presented in Section 5.3.

Within the first three hours post dose, the values of $\Delta\Delta QT_c$ increase, as shown in Figure 8.2. This is highly correlated to the increasing blood plasma curve which is drawn in green in Figure 8.2. A slow decrease in $\Delta\Delta QT_c$ follows during the next four to five hours. However, the curve remains elevated during the rest of the measurement. It can also be seen, that the plasma concentration albeit decreasing continuously, does not reach the baseline until the end of the measurement. In total, a prolongation of the corrected QT interval by Moxifloxacin of about 12 ms can be observed after the administration of Moxifloxacin in THEW tQT_{II}-study. A high correlation between QT prolongation and the blood plasma concentration is seen.

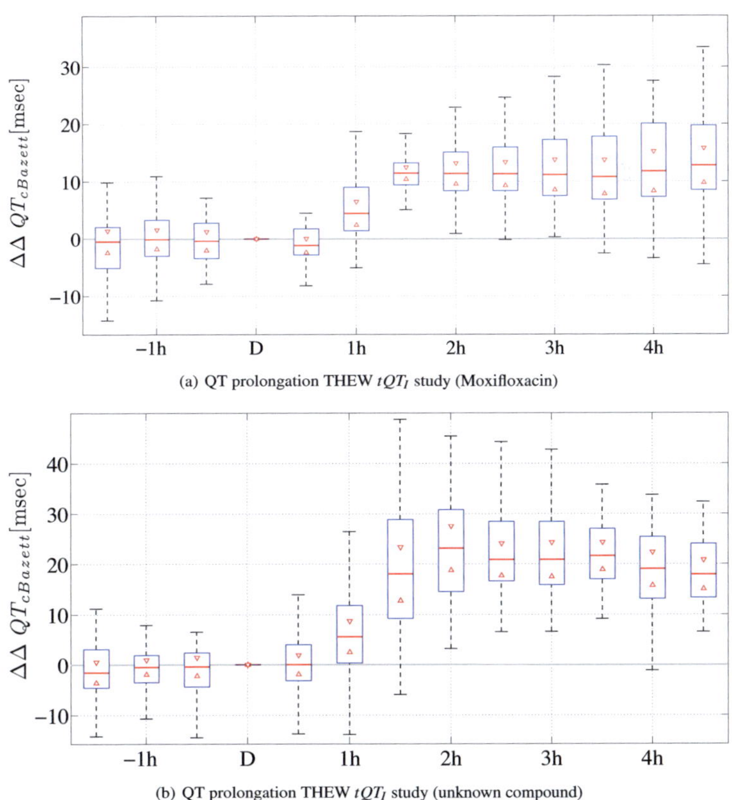

(a) QT prolongation THEW tQT_I study (Moxifloxacin)

(b) QT prolongation THEW tQT_I study (unknown compound)

Fig. 8.3. QT prolongation due to Moxifloxacin and an unknown compound in the clinical trial THEW tQT_I-study. The boxplot include data of 34 evaluated participants of the study. For both, Moxifloxacin and the unknown compound, a significant increase in the corrected QT interval can be observed post dose.

Figure 8.3(a) shows the prolongation of the corrected QT interval in THEW tQT_I-study due to Moxifloxacin. The study is designed with three measurements for each of the 35 participants. One with a placebo, one with Moxifloxacin and one with an unknown compound. The data were evaluated until four and a half hour post dose. 34 participants of the study have been taken into account in this research project.

The time course of $\Delta\Delta QT_c$ is more or less constant pre dose. In the first two hours post dose, a strong increase can be observed, while the time course of $\Delta\Delta QT_c$ is more or less constant for the next hours. The maximum prolongation of the corrected QT interval due to Moxifloxacin is again about 12 ms. As expected, effects of Moxifloxacin induced QT prolongation are quite similar for both THEW studies.

Figure 8.3(b) shows the prolongation of QT_c by an unknown compound. The characteristics of the boxplot are similar to the one for Moxifloxacin. A significant difference is observed for the amount of prolongation. The unknown compound increases the corrected QT interval up to about 23 ms.

8.2.5 Discussion

The work flow of delineating ECG data in the way described in Chapter 7 leads to information, which can be used to detect small drug induced QT_c prolongations in Holter ECG data. The drug induced prolongations seen in the THEW QT_I study are highly correlated to the course of the blood plasma concentration. Thus, a clear relationship between the prolongation and the medication can be assumed. Altogether, QT_c prolongation of both Moxifloxacin studies and also for the UnC study showed similar results as reported in the literature regarding course and maximum mean values [78, 81].

Based on the same data, Darpo et al. [81] show ways to improve the accuracy of precise QT interval measurements in their report, "Improving the precision of QT measurements". The output values of mean QT_c prolongation are very similar to the ones presented in this research work. However, they used the formula of Fridericia to correct QT and the number of participants in tQT_{II}-study is different. The way of data analysis differs, as in [81], only parts of the signal were evaluated, while in this work the whole Holter ECG recordings have been taken into account. The focus of this work is on a general improvement of QT measurements by enhancing the method of T wave end detection. A total delineation has been made and every single QT interval in the ECG has been taken into account. This seems to increase the variance of the results, but guarantees that the QT interval detection has a reliable quality, which is high enough to detect even very small QT_c prolongation. To decrease the standard deviation of the output, an approach based on windowing, beat selection and manual review by an ECG analyst as described in [81] should be used. Nevertheless, concentrating on short parts of the ECG is not reasonable to test the quality of T wave end delineation.

8.3 Drug Induced Morphology Changes of the T Wave

Although a drug induced QT prolongation has been identified, an arrhythmogenic modification of the repolarisation process in the heart is not assured. TdP have been identified to be in some way related to a prolongation of the QT interval, but only few patients with prolonged QT interval suffered from tachyarrhythmias or even TdP. However, a qualitative relationship between the risk of TdP and a QT_c prolongation is obvious [91].

The genesis of the T wave is the result of a gradient of action potentials (AP) of cardiac cells. This gradient comes from the heterogeneity of the AP morphologies and the activation sequences of the cells across entire the heart [103, 104, 105]. Keeping this in mind, an analysis of the T wave morphology to detect changes in the repolarisation process is reasonable. This should also provide new information on the effects of drugs on the complex repolarisation process of the heart.

In this research work, several descriptors based on the T wave morphology have been analysed regarding their sensibility to drug related changes. For this purpose, ECG data of THEW studies tQT_I and tQT_{II} has been investigated. The descriptors have already been introduced in Section 7.5.1.2.

8.3.1 Methods

The ECG data has been delineated and time series have been generated using BSAT. To visualize drug induced repolarisation changes, the morphology of the T wave is analysed. The effects caused by drugs can be observed within several minutes or hours. Thus, instead of applying beat to beat analysis, the observation of longer intervals is reasonable. For Moxifloxacin studies, intervals of 10 minutes have been chosen. This seemed to be a good compromise between having a satisfactory number of T waves and a relative constant drug influence for all waves in the interval. Signal processing as described in Section 7.5.2.3 was done to get a representative T wave of the interval. The mean value determination leads to a reduction of white noise on the wave. Artefacts, not detected in the outlier detection routine, will have a small impact on the resulting mean wave.

All descriptors are computed for each mean wave. The intervals of both recordings (drug and placebo) were synchronised with dose time and thus the recordings can easily be compared. A $\Delta\Delta$ analysis is performed and the values of all descriptors are compared in the same way as described in Section 8.2.2 for the QT_c intervals.

8.3.2 Results of THEW tQT$_{II}$-Study

To evaluate the influence of the drug Moxifloxacin, 57 subjects of the tQT_{II}-study have been analysed. The results can be seen in the boxplots of the following figures.

Regardless of the official numbering of the THEW studies, this section begins with tQT_{II}-study because of the blood plasma concentration measurement. This measurement is only available in the THEW tQT_{II}-study. The author suggests that a comparison to the blood plasma concentration helps to understand the course of the descriptors and allows a more intuitive interpretation of the results.

Descriptors based on Statistical Values

In Figure 8.4(a), expected value μ' of the T wave is shown. The value of μ' seems to be unaffected by the drug contrary to standard deviation σ', which is increased by a maximum of about 1 ms, three hours post dose. This is illustrated in Figure 8.4(b). The increase is correlated to the blood plasma concentration of Moxifloxacin in the first hours, but remains at a relative high level until the end of the recording. An increased standard deviation represents a broadened T wave.

In Figure 8.5, the statistical values of skewness γ_1' and kurtosis β_2' are shown. Skewness and kurtosis are both affected by the drug. While the values of γ_1' are increased post dose, the values of β_2' are decreased post dose. The values of both descriptors are significantly changed post dose in relation to pre dose baseline. The variance of the descriptors within one time interval is increased 5 to 14 hours post dose. A lower value of β_2' is again an indicator for a broader T wave. Increased values of γ_1' represent a change in the symmetry of the wave.

Summarizing, a correlation between the statistical values of the T wave, except μ' and the blood plasma concentration of Moxifloxacin is observed.

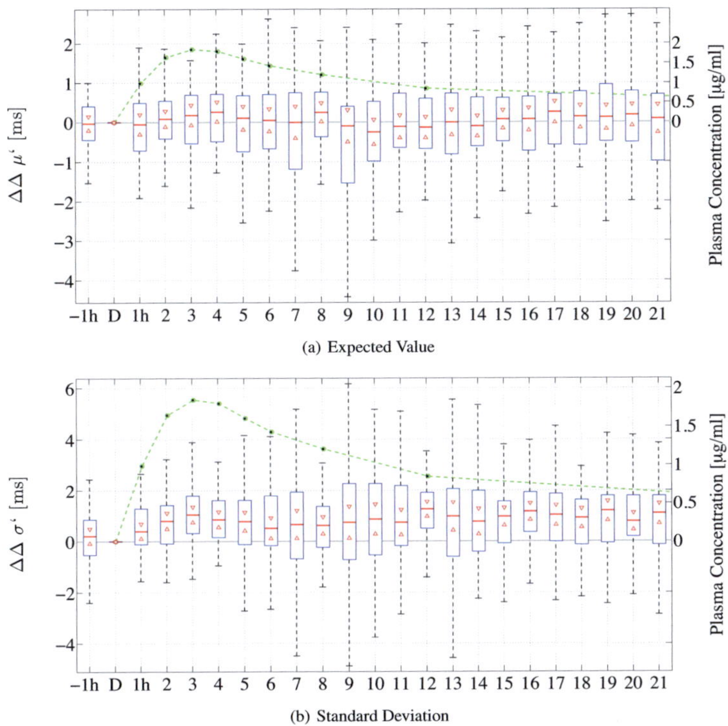

Fig. 8.4. Moxifloxacin induced changes of a) expected value and b) standard deviation of 57 subjects of *THEW tQT$_{II}$* study. Interval length is 10 minutes. Only the first interval of every hour is shown. Time course of the mean blood plasma concentration is shown in green. Time of dose is marked as D.

Descriptors Based on the Slope of the T wave

The drug induced influence on the T wave descriptors α'_I, α'_{II} and height ν are content of Figure 8.6. The maximum gradient on the left half of the T wave, α'_1 is shown in Figure 8.6(a). This descriptor is increased by Moxifloxacin, which indicates a steeper slope. In Figure 8.6(b) the values of α'_2 are decreased, which indicates a steeper slope for the falling edge[2].

Differences of the height of the T wave ν are shown in the lower plot of Figure 8.6(c) as percentage deviation of the T wave amplitude at dose time. Two hours post dose, the values are decreased. Only a small correlation to the plasma con-

---

[2] The falling wave has negative gradient. Thus, a steeper slope leads to an increase in the $\Delta\Delta$-plot.

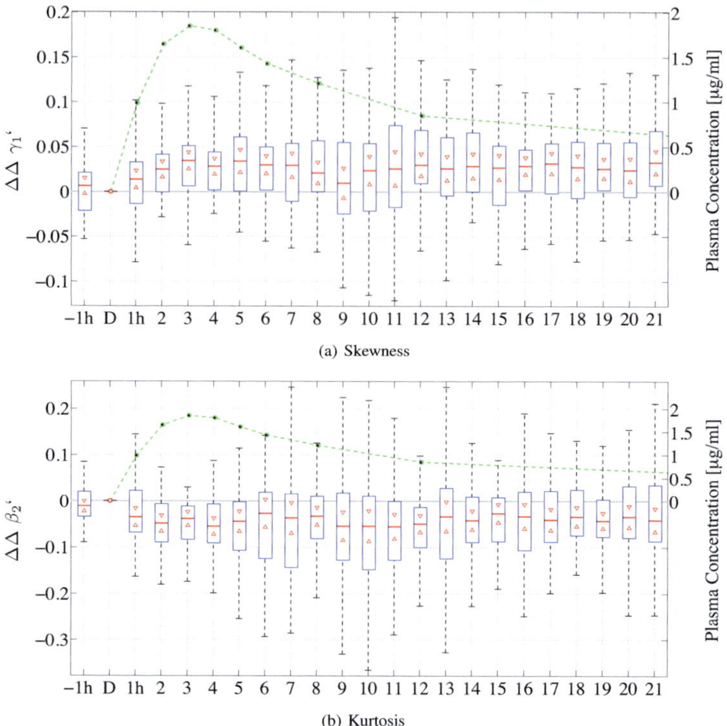

(a) Skewness

(b) Kurtosis

Fig. 8.5. Moxifloxacin induced changes of a) skewness and b) kurtosis of 57 subjects of *THEW tQT$_{II}$* study. Interval length is 10 minutes. Only the first interval of every hour is shown. Time course of the mean blood plasma concentration is pictured in green. Time of dose is marked as D.

centration can be observed in this study. However, a step in the $\Delta\Delta$-plot is clearly seen post dose.

Time Based Descriptors

Content of Figure 8.7 are three time dependent values of the T wave. The descriptors are described in Equation 7.82 to 7.84. Figure 7.34 illustrates the time intervals of the T wave.

δ_l in Figure 8.7(a), describing the length of the right half of the T wave, is increased by Moxifloxacin with a maximum of about 4 ms. The time, the T wave is decreasing, is longer under the influence of Moxifloxacin. Also the duration of the whole T wave is increased by Moxifloxacin as shown in Figure 8.7(b). The

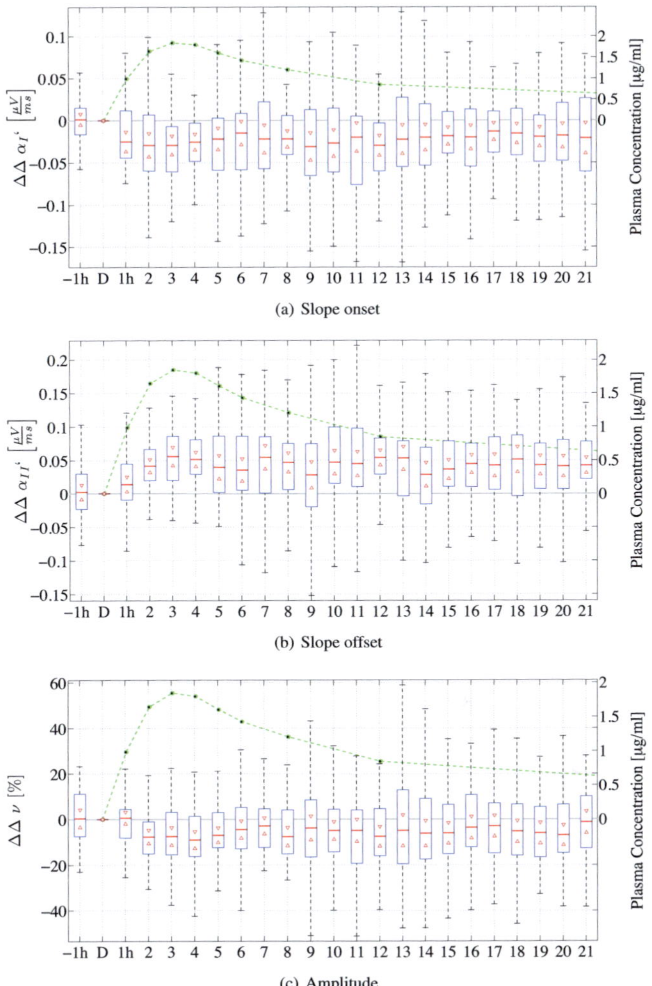

Fig. 8.6. Moxifloxacin induced changes of a) slope onset, b) slope offset and c) amplitude of 57 subjects of *THEW tQT_{II}* study. Interval length is 10 minutes. Only the first interval of every hour is shown. Time course of the mean blood plasma concentration is pictured in green. Time of dose is marked as D. Amplitude values are presented as percentage deviation to the T wave amplitude at dose time.

maximum median value lies at about 5 ms. The time between the inflexion points of the T wave increases by on average 2.5 ms due to the medication. For all three values, a high correlation to the blood plasma concentration of the compound can be observed.

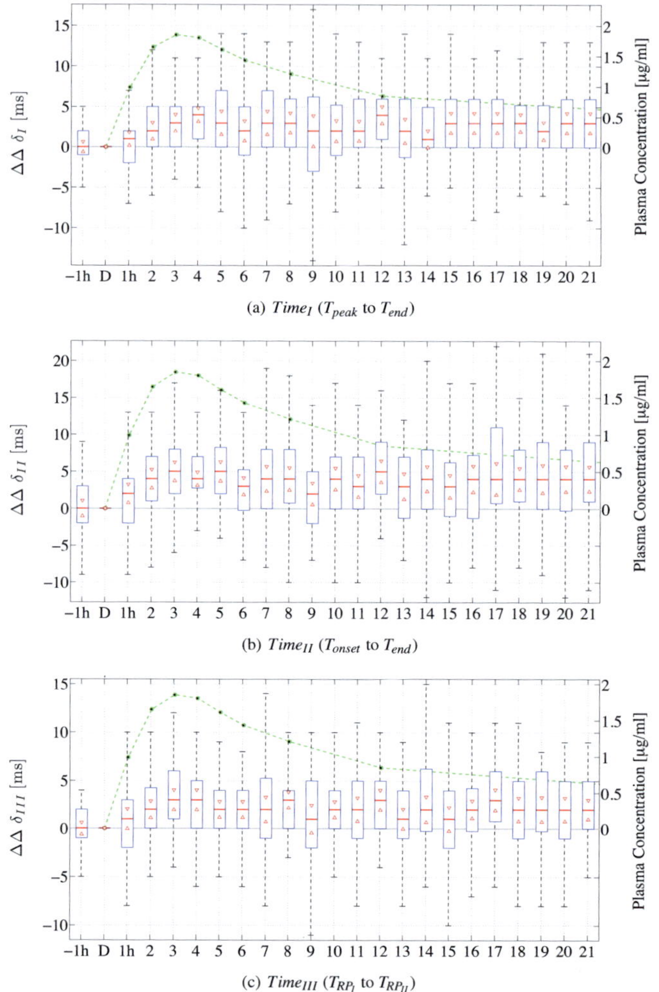

(a) $Time_I$ (T_{peak} to T_{end})

(b) $Time_{II}$ (T_{onset} to T_{end})

(c) $Time_{III}$ (T_{RP_I} to $T_{RP_{II}}$)

Fig. 8.7. Moxifloxacin induced changes of three time intervals of 57 subjects of *THEW tQT_II* study. Interval length is 10 minutes. Only the first interval of every hour is shown. Time course of the mean blood plasma concentration is pictured in green. Time of dose is marked as D.

Area Based Descriptors

The last set of descriptors of the T wave, which have been investigated in this research, are based on the area of the T wave. These descriptors have been calculated by Equation 7.93. An illustration can be found in Figure 7.34.

Figure 8.8 shows the results of this area based descriptors. The impact of Moxifloxacin on all of the three descriptors seems to be small in relation to their variance. Compared to the boxplots of previously shown descriptors, the effect is not significant. Nevertheless, a small decrease in all values about 3.5 hours post dose is discernible.

8.3.3 Results of THEW tQT$_I$-Study

Moxifloxacin

The ECG data of THEW tQT_I-study has been analysed four hours post dose. Same descriptors as previously shown for study *II* have been investigated. For clarity reasons, only the boxplots of THEW tQT_{II}-study have been shown in this section. The boxplots of all descriptors for THEW tQT_I-study can be found in the Appendix A.2. Figure A.1 to A.13 show the influence of Moxifloxacin, while Figure A.14 to A.26 show the effects of an unknown compound.

The effects caused by Moxifloxacin in THEW tQT_I-study and the results of THEW tQT_{II}-study are alike. Only a small variance can be observed for the height of the T wave. While the decrease in v in THEW tQT_{II}-study was small, a reduction of the descriptor v in THEW tQT_I-study is hardly observed.

Time dependent parameters are changed in THEW tQT_{II}-study with a maximum median of less than 5 ms. In THEW tQT_I-study, the prolongation of the time intervals can only be sensed. Caused by the smaller sample rate of THEW tQT_I-study, the values can only change in discrete multiples of 5 ms. However, a Moxifloxacin induced prolongation can be assumed.

Again, all area based values seem to be unaffected by Moxifloxacin.

Unknown Compound

Beside Moxifloxacin, an unknown compound was administered to all subjects in THEW tQT_I-study. Changes of the T wave descriptors due to the unknown compound can be seen in Figure A.14 to A.26. The results are similar to those of Moxifloxacin, but with more pronounced changes. The unknown compound seems to

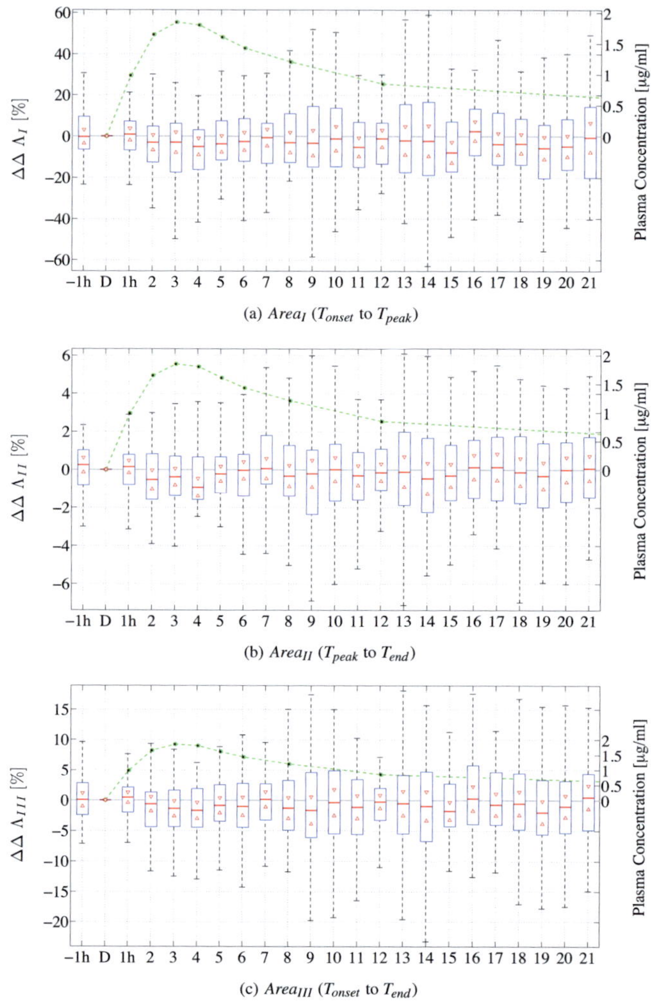

(a) $Area_I$ (T_{onset} to T_{peak})

(b) $Area_{II}$ (T_{peak} to T_{end})

(c) $Area_{III}$ (T_{onset} to T_{end})

Fig. 8.8. Moxifloxacin induced changes of wave areas I to III of 57 subjects of *THEW tQT_{II}* study. Interval length is 10 minutes. Only the first interval of every hour is shown. Time course of the mean blood plasma concentration (green). Time of dose is marked as D.

influence the T wave in a similar way as Moxifloxacin. While the trend for all descriptors is similar, the effects seems to be amplified by the unknown compound, compared to Moxifloxacin. Table 8.1 shows the maximum median values for both studies and both compounds. Altogether, descriptors showed the same trend in all studies. The unknown compound seems to alter the T wave's height and all area based descriptors, which was not as clearly seen in case of Moxifloxacin. Nevertheless, the difference is quite small and the modifications of the descriptors are not as pronounced as in case of e.g. γ_1 or β_2.

Table 8.1. Maximum Median values of all Descriptors within the first 4.5 hours post dose. Unknown compound (UnC), Moxifloxacin (Mox.)

Descriptor	THEW tQT_I				THEW tQT_{II}	
	UnC	Time [Min]	Mox.	Time [Min]	Mox.	Time [Min]
$QT_{cBazett}$	23.11 ms	120	12.72 ms	270	13.06 ms	240
μ'	0.49 ms	10	0.53 ms	230	0.26 ms	240
σ'	2.56 ms	180	1.17 ms	120	1.04 ms	180
γ_1'	0.1	110	0.06	230	0.03	180
β_2'	-0.17	120	-0.08	240	-0.05	240
α_I'	-0.14 $\frac{\mu V}{ms}$	160	-0.07 $\frac{\mu V}{ms}$	160	-0.03 $\frac{\mu V}{ms}$	180
α_{II}'	0.19 $\frac{\mu V}{ms}$	100	0.11 $\frac{\mu V}{ms}$	230	0.06 $\frac{\mu V}{ms}$	180
v	-11.38 %	170	-8.82 %	220	-9.09 %	240
δ_I	5 ms	70	5 ms	110	4 ms	240
δ_{II}	10 ms	100	5 ms	100	5 ms	180
δ_{III}	7.5 ms	140	5 ms	100	3 ms	180
Λ_I	13.78 %	210	6.79 %	120	-4.96 %	240
Λ_{II}	0.81 %	210	0.45 %	80	-0.93 %	240
Λ_{III}	2.55 %	210	1.05 %	250	-1.7 %	240

8.3.4 Drug Induced T Wave Modification

Figure 8.9 illustrates the modification of the T waves of one participant of the THEW tQT_I-study. Mean T waves of the first 10 minute interval of every hour have been displayed for placebo 8.9(a) and Moxifloxacin 8.9(b). The colour of the wave indicates the time when the wave was recorded, starting from the beginning of the record. T waves of both recording days show a moderate variance during the day. Significant differences cannot be observed. To illustrate changes, differences between the waves of both days have been calculated in Figure 8.9(c). To

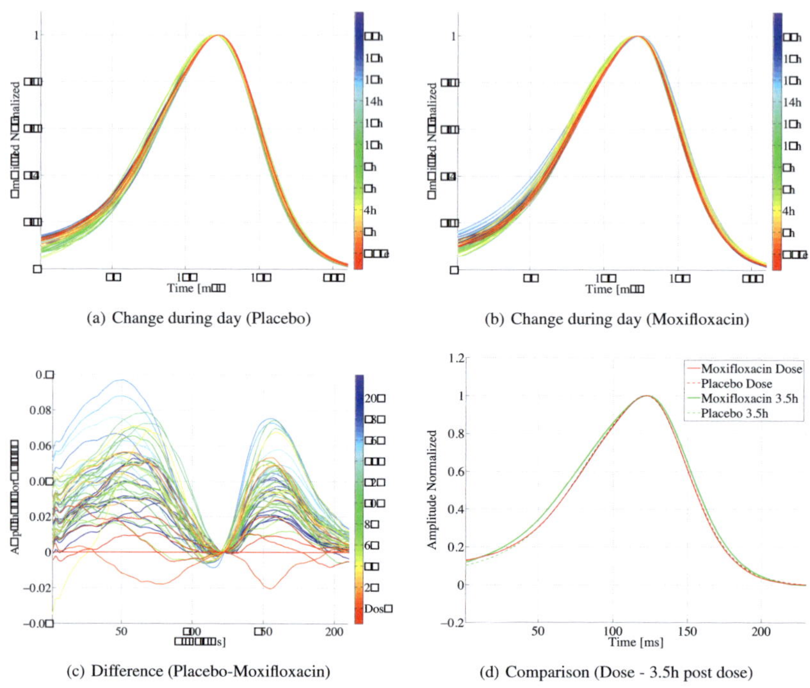

(a) Change during day (Placebo)

(b) Change during day (Moxifloxacin)

(c) Difference (Placebo-Moxifloxacin)

(d) Comparison (Dose - 3.5h post dose)

Fig. 8.9. Changes of the T wave due to Moxifloxacin in one participant. a) Alteration of the T wave during a day with placebo. A clear modification can be seen in the area of T_{onset} and at the increasing and decreasing slope of the wave. b) Alteration of the T wave during a day with a 400 mg Moxifloxacin dose. Changes are similar to those of the placebo day. c) Difference between a) and b). The T wave differences at the dose time have been subtracted. d) Comparison of the T wave at the beginning of the record (green) and 3.5 hours post dose (red). The dashed lines belong to the placebo record. Broadening of the T wave can be observed 3.5 hours after the dose of Moxifloxacin. All waves are aligned with T_{peak}.

prevent inter-day variabilities, the last wave pre dose has been subtracted from all waves of the corresponding day. It can be observed, that changes in the first half of the T wave are more pronounced than those at the right half of the wave. Taking colours into account, an increase in the differences can be identified post dose. In the last few hours of the recording, the differences decrease again (dark blue lines). Taking a look at the wave at the beginning of both recordings and 3.5 hours post dose, as shown in Figure 8.9(d), a clear broadening of the T wave can be observed (green and red solid line). T waves at the beginning of the recording (green and red dashed lines) are very similar to the 3.5 hour post dose wave of the placebo recording (green solid line). Major changes can be observed during the first half of the T wave.

8.3.5 Statistical Evaluation

To test the statistical significance of the parameters of THEW-studies tQT_I and tQT_{II}, three statistical tests have been performed. The samples of the tests contain values of one 10 minutes interval each. Always two sample populations, taken from corresponding 10 minutes interval of placebo and compound measurement, have been tested against equal means. To eliminate the effects of changes between both measurement days, values of descriptors at dose time have been subtracted from all intervals, (equal to the $\Delta\Delta$ analysis).

Tests have been made for all intervals, but for clarity reasons only three representative time intervals have been selected to be presented in the table. To be able to compare all THEW studies and compounds, equal time intervals have been chosen. First group comprise data of the interval 10 minutes pre dose. Group two and three consists of data of intervals 90- and 210 minutes post dose. All tables present p-values as percentage, showing the probability of making a mistake if the null hypothesis H_0 of the statistical test is rejected.

The first column of each tested interval group shows the p_s-values of a sign test. Second column results from a Wilcoxon signed-rank test. Values are named as p_w. Third column represents p_t-values returned by a statistical t-test.

All performed statistical tests, investigate the null hypothesis of equal mean or medians between both samples. The null hypothesis is rejected, if a significance level of $\alpha_{0.05}$ is reached[3]. To get a direct impression of the test results, all cells containing p-values lying below significance level have been coloured green.

THEW tQT_{II}-study, Moxifloxacin

Table 8.2 shows the results of the THEW tQT_{II}-study. Pre dose, significant differences between the samples cannot be observed in any of the descriptors. Post dose, several parameters have significantly been changed. The descriptors seem to be affected by Moxifloxacin, since their values change over time. While many p values are below significance level 90 minutes post dose, the same descriptors have even smaller p-values 210 minutes post dose. As seen in the boxplots, modifications of the T wave caused by Moxifloxacin take some hours to fully develop. This leads to decreasing p-values over time.

---

[3] p-values lying below 5%.

Table 8.2. p-values of different significance tests for all descriptors in **THEW** tQT_{II}**-study** for **Moxifloxacin**. Significance was tested for three 10 minute intervals, one pre dose, one 90 and one 210 minutes post dose. p-values represent the error probability if the null-hypothesis of equal mean or median is rejected. p_s are the p-values of a signum test, p_w are the p-values of a Wilcoxon signed rank test and p_t represent the p-values of a student t-test.

Descriptor	Pre Dose (-10 Min)			Post Dose I (+90 Min)			Post Dose II (+210 Min)		
	p_s [%]	p_w [%]	p_t [%]	p_s [%]	p_w [%]	p_t [%]	p_s [%]	p_w [%]	p_t [%]
μ	28.9	24.4	33.5	100	97.8	95.1	59.7	87.7	95
σ'	79.1	64.8	41.4	0.8	0.1	1.6	< 0.1	< 0.1	< 0.1
γ_1'	6.3	21.1	81.6	3.3	0.6	2.2	< 0.1	< 0.1	0.4
β_2'	79.1	38	93.6	1.6	2.2	2.2	0.1	< 0.1	0.1
α_I	42.7	82.7	48.3	0.8	0.1	6	< 0.1	< 0.1	0.1
α_{II}	59.7	47.2	78.5	< 0.1	< 0.1	1.6	< 0.1	< 0.1	< 0.1
ν	100	85.2	73.7	42.7	16.8	64	11.1	3.8	5.5
δ_I	88	91.1	41.6	0.5	0.2	68.3	< 0.1	< 0.1	0.5
δ_{II}	64.4	50.6	46	0.2	0.2	3.7	< 0.1	< 0.1	< 0.1
δ_{III}	65.2	38.5	58.5	0.1	< 0.1	< 0.1	< 0.1	< 0.1	< 0.1
Λ_I	59.7	95.2	76.6	42.7	71.2	50.4	79.1	74.8	17.6
Λ_{II}	59.7	49.7	40.6	100	97.8	65.9	11.1	58.1	39.3
Λ_{III}	42.7	88.9	61.3	100	98.4	54.6	79.1	90.8	22.5

Significantly changed indicators of a Moxifloxacin induced T wave modification seem to be statistical values like σ', γ_1' and β_2'. Further on, gradients in the inflexion points α_I' and α_{II}', as well as all time based descriptors δ also show statistical significance.

However, expected value, amplitude and area based descriptors show no significant change post dose. Large differences between the tests cannot be observed. As expected, t-test is more conservative than sign and Wilcoxon test. 210 minutes post dose all tests present similar results.

To get an impression of the test results of all 10 minute intervals of the THEW tQT_{II}-study, Figure 8.10 shows a bar diagram of the results of the t-test for some of the descriptors. Red bars indicate intervals having a test result above significance level, while green bars indicate intervals with test results below significance level. In case of statistical significance, the length of the bar indicates the value of the test. Not significant intervals have been drawn with constant length regardless of the value for clarity reasons.

Pre dose, nearly all intervals have red bars. A significant change of the descriptors pre dose is not observed. Post dose, several descriptors show notably significant changes. Specially between two and four hours post dose, most intervals are significantly changed. Influence of the compound seems to be different in duration for the descriptors. Some of them show low p-values during the entire measurement, as e.g. δ_{III}, while others, like e.g. β_2', have significantly changed intervals mainly few hours post dose.

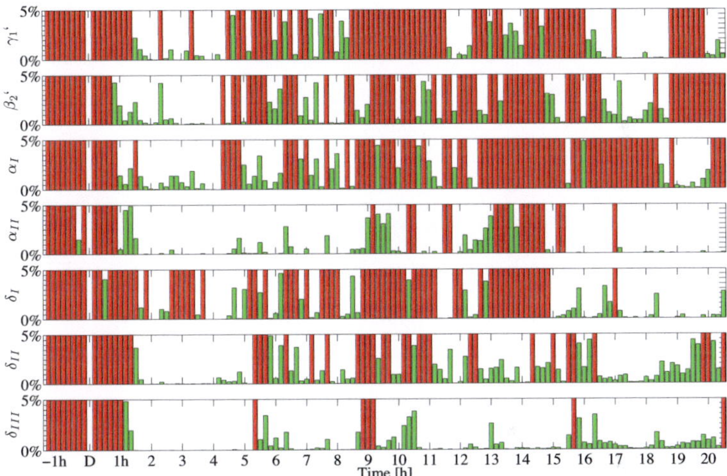

Fig. 8.10. Results of student's t-test for the **Moxifloxacin** study **THEW** tQT_{II}. Red bars illustrate p-values of the t-test higher than significance level of 0.05. Green bars show p-values below significance level of 0.05. It can be seen that pre dose nearly all descriptors show no statistical significance for the difference between placebo and Moxifloxacin day. Post dose, a significant difference can be observed for all descriptors. The number of intervals having a significant difference between both recordings varies. It can be observed that in the interval between 2 to 4 hours post dose, most intervals have a significance below 5%.

THEW tQT_I-study, Moxifloxacin

In Table 8.3, the results of the test for the THEW tQT_I-study and the compound Moxifloxacin are presented. Time intervals are equal to Table 8.2. All descriptors show no significant change pre dose. 90 minutes post dose, Moxifloxacin leads to no significant changes for expected values, amplitude and area based descriptors of the T wave. Results for descriptor δ_I are different to THEW tQT_{II}-study, as all

Table 8.3. p-values of different significance tests for all descriptors of the **THEW tQT_I-study** for **Moxifloxacin**. Significance was tested for three 10 minute intervals, one pre dose, one 90 and one 210 minutes post dose. p-values represent the error probability if the null-hypothesis of equal mean or median is rejected. p_s are the p-values of a signum test, p_w are the p-values of a Wilcoxon signed rank test and p_t represent the p-values of a student t-test.

Descriptor	Pre Dose (-10 Min)			Post Dose I (+90 Min)			Post Dose II (+210 Min)		
	p_s [%]	p_w [%]	p_t [%]	p_s [%]	p_w [%]	p_t [%]	p_s [%]	p_w [%]	p_t [%]
μ	39.2	71.3	96.2	60.8	57.8	32.6	86.4	81.7	40.4
$\sigma^{\prime}$	86.4	72.6	41.4	0.3	0.1	0.2	0.1	< 0.1	0.1
$\gamma_1^{\prime}$	39.2	44.7	79.5	0.9	2.7	10.9	2.4	1.9	4.1
$\beta_2^{\prime}$	86.4	52.1	50.3	12.1	2.1	1.8	0.3	0.1	0.2
α_I	39.2	31.7	51.2	0.1	< 0.1	0.1	0.1	< 0.1	< 0.1
α_{II}	86.4	59	50.6	0.1	0.1	0.1	< 0.1	< 0.1	< 0.1
v	39.2	83.1	75.6	86.4	62.6	52.2	60.8	36	31.6
δ_I	18	33.7	23.1	13.4	22.4	21.4	0.1	0.2	0.1
δ_{II}	26.7	22.7	16	3.5	2.5	6.5	0.2	0.3	1.9
δ_{III}	52.3	40	37	3.5	3.6	16.7	< 0.1	0.1	0.8
Λ_I	22.9	61.4	86.6	12.1	20.3	29.8	60.8	61.4	37.4
Λ_{II}	86.4	85.8	59.5	86.4	83.1	92.8	86.4	75.2	73.7
Λ_{III}	39.2	71.3	91.5	39.2	43.7	48.1	86.4	73.9	48.2

p-values lie above significance level of 5%. t-test of the time dependent descriptors δ_{II} and δ_{III} also fails.

Third group 210 minutes post dose shows similar results as in the tQT_{II}-study.

THEW tQT_I-study, Unknown Compound

Table 8.3.5 represents the results of the tQT_I-study for an unknown compound. Again no significant changes in the descriptors can be observed pre dose. Post dose, no significance can be seen for changes in expected values and T wave amplitude. Changes of all time dependent descriptors are significant in all three tests. Contrary to Moxifloxacin, the unknown compound changes the area based descriptors significantly within 210 minutes post dose. Λ_I seams to be significantly changed already 90 minutes post dose.

In contrast to all results of Moxifloxacin, t-test shows no significant change for the standard deviation and kurtosis of the T wave. Sign and Wilcoxon signed rank test return results of high significance 90 and 210 minutes post dose for these descriptors.

Table 8.4. *p*-values of different significance tests for all descriptors of the **THEW** tQT_l-**study** for the **unknown compound**. Significance was tested for three 10 minute intervals, one pre dose, one 90 and one 210 minutes post dose. *p*-values represent the error probability if the null-hypothesis of equal mean or median is rejected. p_s are the *p*-values of a signum test, p_w are the *p*-values of a Wilcoxon signed rank test and p_t represent the *p*-values of a student *t*-test.

Descriptor	Pre Dose (-10 Min)			Post Dose I (+90 Min)			Post Dose II (+210 Min)		
	p_s [%]	p_w [%]	p_t [%]	p_s [%]	p_w [%]	p_t [%]	p_s [%]	p_w [%]	p_t [%]
μ	39.2	81.7	33.4	39.2	43.7	38.9	86.4	60.2	32.5
σ'	60.8	65.1	32.4	< 0.1	< 0.1	28.6	< 0.1	< 0.1	29.3
γ_1'	100	62.6	28.8	< 0.1	< 0.1	< 0.1	< 0.1	< 0.1	< 0.1
β_2'	60.8	68.8	24.7	< 0.1	< 0.1	57.9	< 0.1	< 0.1	92.4
α_I	39.2	93.9	37.8	< 0.1	< 0.1	< 0.1	< 0.1	< 0.1	< 0.1
α_{II}	39.2	25.6	25.9	< 0.1	< 0.1	< 0.1	< 0.1	< 0.1	< 0.1
v	12.1	87.1	34.4	60.8	9.9	55	86.4	88.4	36.7
δ_I	79.1	79.1	60.1	< 0.1	< 0.1	< 0.1	< 0.1	< 0.1	< 0.1
δ_{II}	48.1	24.5	100	< 0.1	0.1	1.2	< 0.1	0.1	0.7
δ_{III}	35.9	27.7	94.3	< 0.1	< 0.1	0.3	< 0.1	< 0.1	0.1
Λ_I	86.4	80.4	48.2	5.8	2.2	4.1	0.3	< 0.1	0.3
Λ_{II}	12.1	72.6	39.5	39.2	43.7	21.3	5.8	1.3	2.6
Λ_{III}	39.2	87.1	44.9	22.9	7.1	7.9	0.9	0.1	0.6

Summarizing, *p*-values of descriptors affected by the unknown compound are smaller compared to *p*-values of descriptors affected by Moxifloxacin.

8.3.6 Discussion

In this investigation, drug induced repolarisation changes have been analysed using Holter ECG data. New methods of analysing the repolarisation using ECG data and specially considering the T wave, have been applied to two thorough QT studies as they are claimed by the FDA for drug approval [91]. The repolarisation analysis is focused on T wave morphology determining descriptors. These have been compared using ECG measurements of healthy subjects, who were administered either a drug or a placebo.

Results of the $\Delta\Delta$-analysis showed a significant change of several T wave morphology describing parameters. T wave has been changed in duration, skewness and width. Also changes of the maximal gradient of the normalized T waves could be observed and can be ascribed to the drug. To assure the change of the descriptors, three different statistical tests, which showed significant changes of those parameters, have been carried out.

A drug induced change of the amplitude of the T wave (v) was not identified in these studies. The T wave amplitude varies, but a trend which could be assigned to the drug was not observed. This might also be a reason for the poor results of all area based descriptors. A reliable change due to Moxifloxacin was not observed. In case of the unknown compound, the values were significantly changed 210 minutes post dose, while the T wave amplitude showed no significant changes. But, this drug caused a more pronounced modification of all parameters compared to Moxifloxacin. It is conceivable that changes of the T wave morphology exceed the impact of the variance of the amplitude on the area based descriptors. Significant changes of the descriptors α_I, α_{II} and δ_I to δ_{II} can be observed for the unknown compound.

Nevertheless, area based descriptors should not be the first choice for the analysis of the repolarisation process in the heart using surface ECG data.

Time dependent values of the T wave have been considered for all studies of this work. The values of THEW tQT_{II} have a resolution of 1 ms, as the sample rate of the recordings was 1 kHz. For the recordings of THEW tQT_I-study, sample rate was 200 Hz and thus the smallest observable changes is 5 ms. The signals were not sampled up, but this might improve the results for all studies. As the entire analysis involves every heart beat of all ECG recordings, the process of up-sampling leads to a huge increase in computing time. A solution for some of the descriptors, especially time dependent ones, is planned for further research projects.

This discussion is based on the values of the THEW tQT_{II}-study. The results of THEW study tQT_I seem to confirm the former ones, even though they are not as clear owing to the reported sample rate issue.

The time dependent value δ_I gives information on the change of the duration of the T_{peak} to T_{end} region of the T wave. This region seems to be prolonged in maximum about 4 ms for Moxifloxacin. This is not surprising, as an increased QT_c leads to the assumption of a prolonged T wave. On the other hand, the pure change of QT_c does not necessarily lead to a prolonged T wave. If the heterogeneities of the cells in the heart would not be affected, the T wave would probably stay constant in morphology and solely move. The prolongation of δ_I is moreover an indicator that the falling edge of the wave is affected by the drug.

The second time dependent value δ_{II} is a measure of the total duration of the T wave. This value is based on the measurement of T_{onset}, which is less reliable than T_{end} or T_{peak}. Changes in the gradient of the T_{onset} to T_{peak} region affect the detection of T_{onset}. As the gradient in this region is often smaller than the one of the falling edge of the T wave, a variance of this value should be in mind. Nevertheless, a prolongation of the duration of the T wave is clearly seen post dose.

δ_{III} gives information on the slip of the inflexion points of the curve. Information on the shift of every single point is not given. Only changes in distance are measured. The values seem to be prolonged by Moxifloxacin. This is an indicator for a broader T wave. This value returns similar information as the parameter σ', but is, as all time dependent parameters, not affected by the normalisation of the wave.

Two more powerful descriptors for drug induced repolarisation changes are α'_I and α'_{II}. These descriptors are very similar to αL and αR described in the work of Couderc [106], but they differ in the normalisation. In this work, the effects of the variation of the amplitude have been reduced by normalisation. This was also done for the parameters α'_I and α'_{II}. The unit used in this work is equal to the one of αL and αR, but the values are different[4].

Both parameters were significantly changed in all studies and both compounds post dose. The values of α'_I decreased. This means a smaller gradient post dose for the rising slope of the wave. Figure 8.9(d) visualizes that the broader wave causes a slightly decreased gradient between the red and the green curve.

Values of α'_{II} are elevated post dose. This is again due to the broader wave, but because of the negative gradient, the $\Delta\Delta$ values of α'_{II} are increased.

All values used to describe the T wave's morphology resulting from the statistics, have been calculated using the normalised T wave. This is a necessary and important step, which allows the usage of those parameters. In addition, changes in amplitude, which can occur during Holter ECG recordings due to chemical interactions of the electrode-skin coupling or because of a loosening of the electrodes, do not affect those parameters. This makes the results more reliable in clinical praxis. A damping of the values is conceivable, but the results of both studies show, that the effects of Moxifloxacin were not annihilated.

μ' is a parameter, which was not affected by the drugs. This was not surprising, as the T waves have been aligned against each other. Moreover, a static value might

---

[4] During the normalisation process, the amplitude of the wave is changed. Thus α'_I and α'_{II} show the gradient change of the normalized wave. The values are not equal to the maximum gradients of the not normalized wave, but they still describe a change of voltage per time.

be an indicator of high quality data. Changes in the expected value should be moderate, so that they can be related to small changes of the T wave, as e.g. a small drug induced increase at the beginning of the T wave.

The values of σ' increased in good correlation to the blood plasma concentration of Moxifloxacin. This is an indicator that the T wave is broadened by the drug. Couderc et al. came to similar conclusions in several works, as e.g. [106].

If the width of the T wave is increased, the values of kurtosis should decrease, as the wave is expected to get less peaked. The values of β_2' in both THEW studies were decreased post dose. This is congruent to the expectations, but might lead to the conclusion that calculating this value is obvious, as it gives equal information as the standard deviation in this context. The values of kurtosis are more sensitive to changes in the region around the maximum of the wave.

The values of skewness γ_1' were also significantly increased due to Moxifloxacin and the unknown compound. It gives the impression of a smaller decrease of the T wave (for positive T waves) and thus a longer left part of the wave. This could not be observed. Rethinking the measure of skewness leads to the following statements:

- Skewness is a measure of the symmetry of a distribution. The Gaussian distribution has the value zero, as it is symmetrical.
- For positive skewness, values smaller than the distribution's mean value are observed more often than values higher than the mean.
- The peak of the distribution curve is on the left side from distribution's mean value for positive skewness.

Values at the beginning (left side) of the wave are affected by the drug and do not necessarily start at zero. This is different to most distributions, which normally begin and end with values near zero. An increase in values smaller than the expected value leads to a positive skewness as heard before. An increase of the T_{onset} to T_{peak} region as observed in Figure 8.9(d) leads to positive values of γ_1'. This value can give information on the modification of the balance between T_{onset} to T_{peak} region and T_{peak} to T_{end} region. Hence, skewness is a powerful parameter for drug induced repolarisation changes.

The reliability of the descriptors was proved and introduced in this section. Dependency on the heart rate of the descriptors was not analysed. To be sure to ascribe the modifications to the compound, without having effects of the heart rate, as in case of the QT interval, a detailed analysis of the effects of the heart rate on the

T wave describing parameters has to be carried out. This is part of the investigations in the following section.

8.4 Heart Rate Dependency

One of the major drawback of analysing the QT interval in ECG data is the fact of heart rate dependency [96, 97, 100, 107]. The dynamics between QT and RR is still in focus of research and a general solution for a mathematical description of the QT/RR relationship in human is not possible, if high accuracy is claimed [102]. Several parameters describing the morphology of the T wave and thus the repolarisation of the heart, have been introduced in previous sections of this work. Heart rate dependency of those descriptors was not in focus yet. In this section, a research work on heart rate dependency of the introduced descriptors is presented.

8.4.1 Methods

To test parameters for heart rate dependency, IBT-Exercise-study has been used. This study involves 10 participants, who underwent a stress test on a bicycle ergometer to change the heart rate in a wide range. ECGs have been recorded with *BSAT-1012*. A detailed description of the data is given in Section 5.5.

Descriptors, which have been identified to significantly change due to Moxifloxacin in both THEW studies, were tested for heart rate dependency. For clarity reasons, area based values are not described here, as a drug induced change has not been observed.

During sporting activity, very fast changes of the heart rate can be observed. A representative T wave resulting from several T waves in a 10 minute interval as used for analysing ECG data of both THEW studies is not reasonable. Hence, parameters have been calculated for every single beat in this test. The values have been compared to the RR interval length of the corresponding heart beat. Signal processing and data export has been carried out as described in Chapter 7.

8.4.2 Results

To illustrate the influence of the heart rate on the descriptors, scatter plots have been drawn from all recordings and all descriptors. Results of two different participants are representatively shown in this section. Figure 8.11 visualizes the measured QT intervals depending on the heart rate of participant P_1 and P_2, respectively. In both figures, often described QT/RR hysteresis [95, 108, 109] is clearly seen. Referring to those images, RR dependency of other descriptors can be compared. Even though both scatter plots of Figure 8.11 show clear hysteresis loops, small distinctions in QT/RR coupling are obvious. Even within one recording, the QT/RR coupling seems to alter.

To illustrate the chronological sequence, markers have been coloured. Starting at dark blue over green, yellow to red, an impression of the time throughout the recording is given. Taking colours into account, time can be considered and small modifications of the hysteresis loops can be seen.

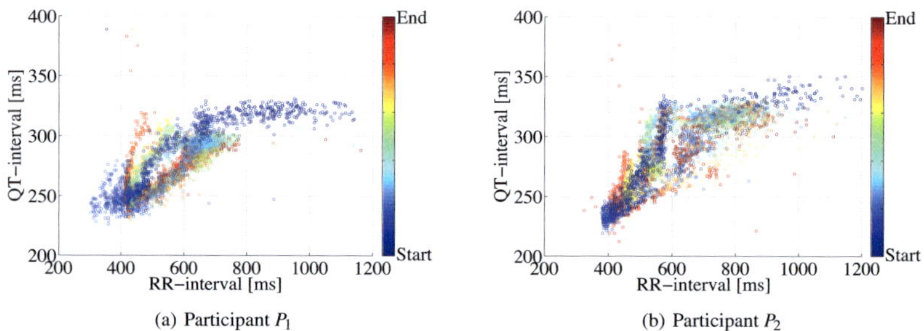

(a) Participant P_1 (b) Participant P_2

Fig. 8.11. Scatter plot: QT- vs. RR interval of participant P_1 and P_2 during stress test. Clear hysteresis in QT/RR coupling can be observed. QT interval for heart rates of about 60 bpm (1000 ms RR interval) lies in the healthy range between 325 ms and 450 ms. Time during recording is illustrated by the colours.

Figure 8.12 and 8.13 show RR interval dependency of 10 descriptors for P_1 and P_2. For all descriptors in both subjects, a clear change of the parameter can be observed for RR intervals below $\sim$700 ms ($\sim$86 bpm). Above 700 ms RR interval parameters remain more or less stable.

Taking the colours into account, it can be seen that changes often start for short RR intervals (high heart rates) and take some more time to get back to baseline compared to the RR intervals. Hence, small hysteresis loops can be observed. The effects slightly differ between the subjects and throughout the parameters. Hysteresis loops of e.g. α'_I and α'_{II} are more pronounced in Figure 8.12 and 8.13 than those of time based parameters δ_I to δ_{III}. σ' and β'_2 show very similar behaviour. γ'_1 has only a small hysteresis for high heart rates. Amplitude seems to have strongest dependency on the heart rate, especially for low heart rates (RR intervals above 700 ms). There, no heart rate dependency was observed for most parameters.

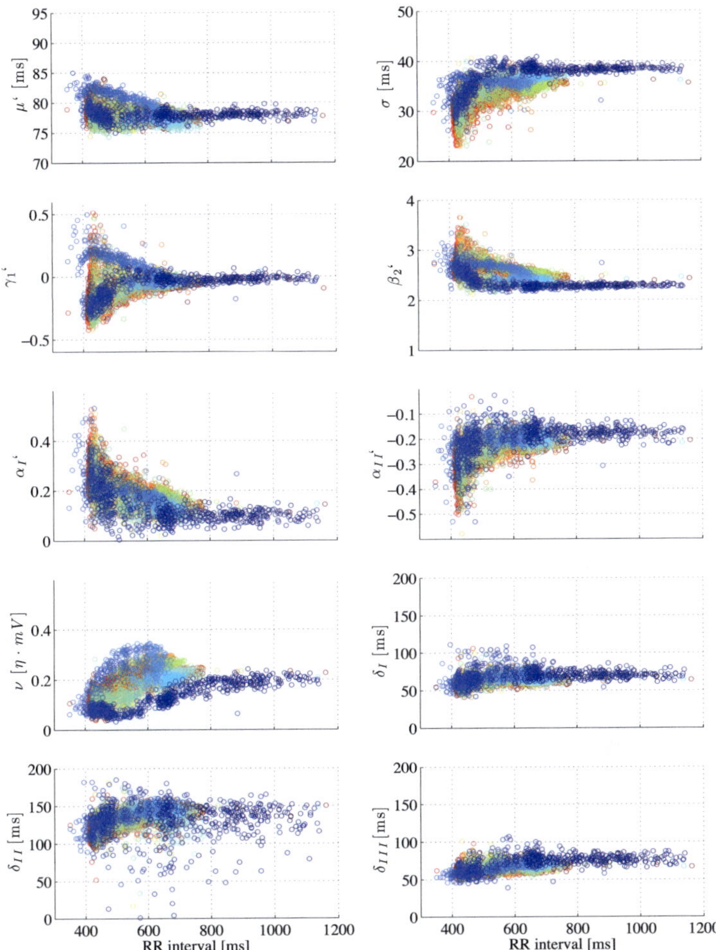

Fig. 8.12. Heart rate dependency of T wave morphology descriptors during stress test of participant P_1. Parameters resulting from statistical measurements as $\mu\text{'}$, σ', γ_1' and β_2' seem to stay stable for RR intervals above 700 ms (86 bpm). Below, a change of the values can be observed. Similar behaviour can be seen for time intervals δ_{I-III}. A moderate dependency of the amplitude of the T wave, ν can be observed for longer RR intervals, while a significant change of the amplitude with genesis of a hysteresis loop is visible for short RR intervals. Slope on α_I and slope off α_{II} show a moderate dependency on the heart rate in an interval between 700 ms and 800 ms RR interval. Below 700 ms RR interval values alter significantly and hysteresis loops can be observed. However, the variance seems to be small for heart rates above 800 ms. Colours of the cycles show time during recording, as illustrated in Figure 8.11.

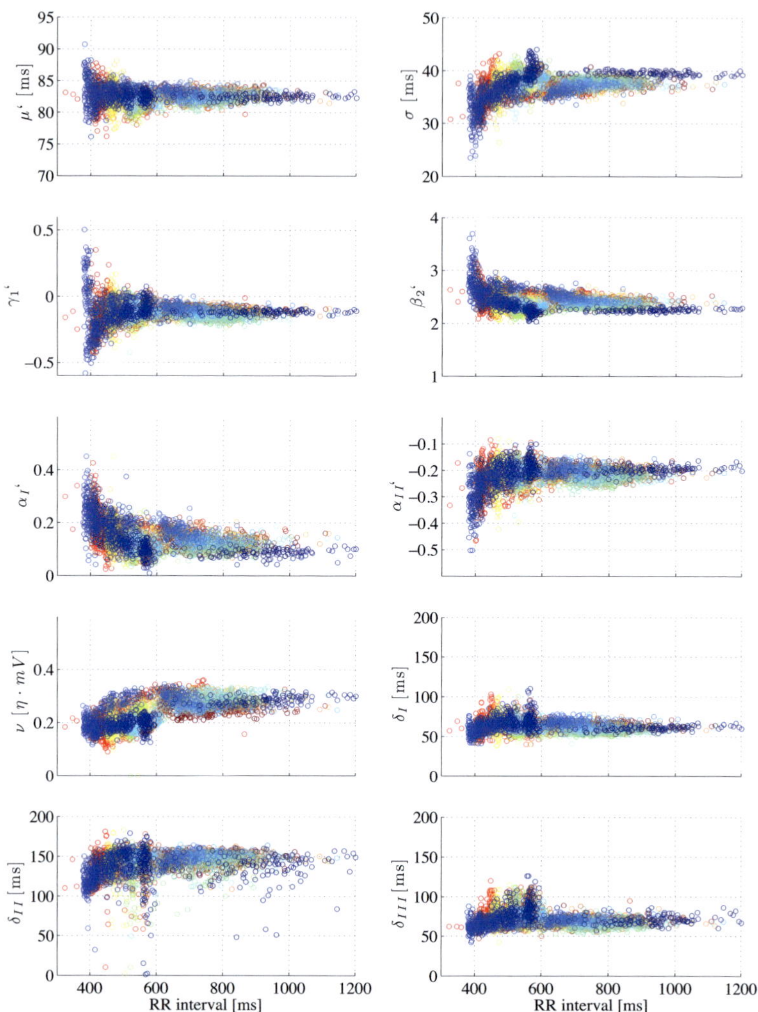

Fig. 8.13. Heart rate dependency of T wave morphology descriptors during stress test of participant P_2. Observations are similar to those of Figure 8.12

Figure 8.14 shows the change of the T waves of P_1 for RR intervals between 400 ms and 1100 ms in an 3D illustration. The plotted waves are generated by calculating

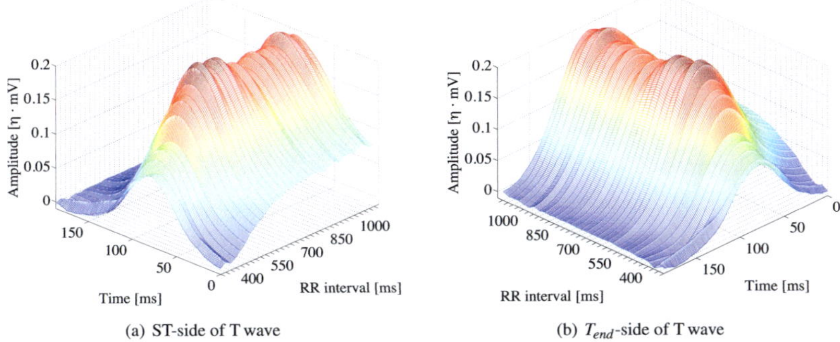

(a) ST-side of T wave   (b) T_{end}-side of T wave

Fig. 8.14. 3D plot of mean T waves for different RR interval lengths

a mean wave of all T waves with similar RR interval length. RR intervals have been binned in intervals of 25 ms. Starting at 400 ms, T waves of all heart beats with an RR interval length between 400 ms and 425 ms have been grouped and the mean wave was calculated for the first wave. Same procedure was carried out for all other RR interval groups.

It is conspicuous that the amplitude is significantly decreased at high heart rates in Figure 8.14. Additionally, a clear decrease at the beginning of the wave can be observed (Figure 8.14(a)). Thus, highest heart rates also have a significant effect on the ST segment in the ECG [110].

8.4.3 Discussion

In this investigation, heart rate dependency of repolarisation describing parameters has been analysed. The influence of 10 of the previously introduced descriptors, depicting the morphology of T wave, have been reviewed on the influence of heart rate. It has been seen, that heart rate changes in a range below about 86 bpm (700 ms RR interval length and more) do not lead to significant changes, whereas a high load due to sporting activity resulting in high heart rates change the descriptor values significantly. This effect is not surprising, as very short RR intervals are intuitively expected to lead to shorter T waves. Values resulting from of the standard deviation or kurtosis as, e.g. σ and β_2, will inevitably change. As well as σ and β_2, time depended parameters δ_I to δ_{III} are constant for heart rates below 86 bpm and decrease for high heart rates.

Expected values and skewness change only for high heart rates. It seems that also the symmetry of the T wave changes under fast heart activity. On the other hand, it can be seen in Figure 8.14 that especially the ST segment of the ECG is changed for high heart rates. This was also assumed to be the reason for changing γ_1' in both THEW studies. Hence, skewness is a marker also for the ST segment change and highest heart rates affect the ST segment.

A decrease at the left part of the T wave necessarily leads to an increase in the expected value, as the T waves have been aligned against each other by minimising a correlation coefficient to a generated mean T wave. Expected value is a parameter, which has to be calculated to be able to derive other parameters as skewness and kurtosis, but it is not as easy to interpret. It is expected to stay stable, if waves have been aligned correctly, but is not independent of changes of the ST segment. The change of μ' becomes comprehensible after viewing Figure 8.14.

γ_1' changes for high heart rates. A change of the increasing or decreasing edge of the T wave is assumed to cause the change . In Figure 8.14 can be seen that indeed the beginning of the T wave is changed for high heart rates. For RR intervals longer than 700 ms, the beginning of the T wave does not significantly alter.

The T wave amplitude seems to be heart rate dependent. Especially in Figure 8.12 related to P_1, a decrease is observed for the dark blue markers. A clear distinct loop can be observed. The T wave amplitude decreases for increased heart rates, which was also observed by Couderc et al. in [111] and Anderson et al. in [112]. In Figure 8.11 a reduction of the QT intervals for decreasing RR intervals can be observed. The decrease of QT rises significantly for RR intervals below 700 ms. This corresponds to the scatter plots of v. Heterogeneities seem to decrease in the heart muscle for highest heart rates, which leads to a smaller T wave.

Changes in amplitude necessarily lead to changes of all area based descriptors. A heart rate dependent T wave might be the reason for the high individual values of amplitude and area based descriptors in both THEW studies introduced in Section 8.2. The authors of [111] and [112] come to equal conclusion for area based descriptors of the T wave.

Recapitulatory, morphology based descriptors of the T wave are heart rate dependent for high heart rates above 86 bpm (RR intervals below 700 ms). For lower heart rates, a significant influence of the heart rate was not confirmed. The only parameter with a light tendency to heart rate dependency was the amplitude. As drug safety studies are mostly based on Holter ECG data without any activity tests, a correction of the morphology describing parameters is not necessary.

On the other hand, information on the relationship between some of the parameters and the heart rate might be interesting for clinical research. Stress test ECG is often used for patients suffering or supposed to suffer from coronary artery disease. Therefore, the relationship between RR and the morphology of the T wave might lead to new and more reliable information on the patients' state of health. J point and ST segment of the ECG are in focus of exercise testing [110]. Changes in these ECG regions will have an effect on the T wave morphology and should be involved for analysis of the repolarisation.

8.5 Conclusion

The observations made in this research let the author come to the conclusion that drug induced repolarisation changes are detectable using the introduced methods. Analysing the morphology of the T wave in the ECG gives more detailed information on the repolarisation process of the ventricles.

Quality of mobile ECG acquisition was significantly improved in the last decades. But a huge amount of information contained in the Holter ECG is not used today, because most signal processing is focused on heart rate variability. This is not necessary, as the quality of the recording is by far good enough for a reliable signal processing on the morphology of the T wave. Holter systems exists in a various number of types, which can be used during the day and even during sporting activities. Using information from those recordings can lead to a significant increase in quality of life for people at high risk for heart attacks. During sporting activity, an analysis of the repolarisation might help to predict a sudden cardiac death within minutes before occurrence. This could help to prevent people from sudden cardiac death in sports.

The analysis of the repolarisation using the ECG might provide new approaches in research to improve therapies and prevention. This additional information of the repolarisation process of the ventricles might also give information on the health state of a person, or help to dose medications in the clinical routine and at home.

9

Analysis of QT/RR Coupling using MVAR-Model Estimation

The QT interval is the most common biomarker used to analyse the ventricular repolarisation based on the surface ECG as reported in Chapter 8. An increase of the heart rate corrected QT interval is assumed to cause arrhythmias such as e.g. TdP. To improve the heart rate correction of the QT interval, the coupling between QT and RR has been widely analysed. QT/RR dynamics seems to be highly different even among healthy people [102]. Therefore, a golden standard for heart rate correction of the QT interval was not found yet [90, 113].

However, changes of the heart leading to modifications of the repolarisation process might also affect the coupling between QT and RR. Hence, analysing the coupling might provide information on the state of the heart. These information might help to asses heart diseases, stress or effects of a medication.

In this chapter, an introduction to the analysis of QT/RR dynamics using a Multi-Variate AutoRegressive model (MVAR) with time series of RR- and QT intervals, is presented.

9.1 Introduction

The use of MVAR models provides new methods for analysing ECG data. The results might help to get information on the physical state of a subject or on dose-effects of drugs.

A two dimensional model is used for estimating parameters describing the relationship between time series of QT and RR. Results are just a sample; some other model settings have been investigated in [10]. However, a clear trend for a model setting was not found. Thus, only one well working model setting is presented for clarity reasons in this chapter.

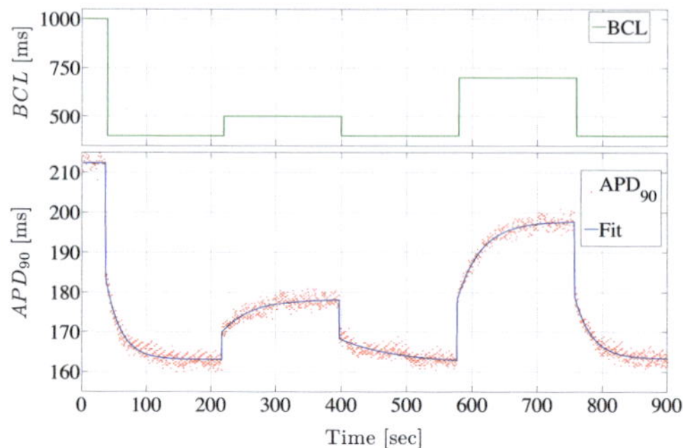

Fig. 9.1. Response of the action potential duration of a heart cell to a step in stimulation frequency.

9.1.1 QT-RR relationship

The dynamics between QT and RR is often discussed, mostly in relation to a heart rate correction formula. The results are quite different, even though step response seems to be similar among different subjects. In 2007 Halamek et al. introduced a publication on the dynamic QT/RR coupling in patients with pacemakers [114]. In this work, participants are stimulated by a special protocol, allowing the heart rate to suddenly change in a significant step. This sudden step of the heart rate was answered by an exponential function of the QT interval. This means, after the step of the heart rate, an adaptation is started to a new QT interval length corresponding to the new RR interval length. This adaptation takes several heart beats and follows an exponential function.

In 2008, the author of this work observed a similar behaviour in myocyte measurements [99, 100, 101]. Myocytes have been stimulated and the action potential duration (APD_{90}) was measured. The dynamics between the basic cycle length (BCL) and the APD_{90} after a step in BCL was similar to the behaviour shown in [114]. Figure 9.1 shows an example of the APD_{90} response after changing BCL in a myocyte measurement.

Of cause, there is a fundamental difference between RR intervals and BCL as well as between the APD_{90} and the QT interval. However, the dynamics should be similar, as the QT interval is a response of all myocytes in the ventricles.

The immediate step of the APD_{90}, following the step of the BCL was not described in [114]. Thus, and to keep the model small, the introduced model does not cover the immediate step of the APD_{90} in this investigations.

The exponential adaptation of the QT interval length to a new RR interval length was the motivation to use a first order MVAR model to estimate parameters describing the dynamics between QT and RR.

9.1.2 Mathematical Description

In this section, some mathematical assumptions on the dependency of a QT interval on previous RR- and QT intervals are made. A step in a RR interval time series, $\boldsymbol{RR}$, is given by a change from RR_A to RR_B. Accordingly, the response of the QT time series $\boldsymbol{QT}$ is from QT_A to QT_B. But, the step in $\boldsymbol{RR}$ is answered by an exponential adaptation of $\boldsymbol{QT}$, as explained before. The adaptation from the 'old' value QT_A to the new value QT_B after a step in $\boldsymbol{RR}$ can be described as:

$$\boldsymbol{QT}(i) = QT_B - (QT_B - QT_A)e^{-\lambda iT} \tag{9.1}$$

with i being the beat number after the step in $\boldsymbol{RR}$ and λ being the rate constant of the adaptation to the new value QT_B. T is the time between two successive heart beats. In these explanations, the time T is constant after the step in $\boldsymbol{RR}$, as $\boldsymbol{RR}$ remains stable on RR_B.

The dependency of a QT interval at heart beat i on a new QT interval at heart beat $i+1$ is now analysed:

$$\Delta\boldsymbol{QT}(i) = \boldsymbol{QT}(i+1) - \boldsymbol{QT}(i) \tag{9.2}$$
$$= QT_B - (QT_B - QT_A)e^{-\lambda(i+1)T} - QT_B + (QT_B - QT_A)e^{-\lambda iT} \tag{9.3}$$
$$= (QT_B - QT_A) \cdot (1 - e^{-\lambda T})e^{-\lambda iT} \tag{9.4}$$

The expression $QT_B - QT_A$ can be replaced according to Equation 9.1 by:

$$QT_B - QT_A = \frac{QT_B - \boldsymbol{QT}(i)}{e^{-\lambda iT}} \tag{9.5}$$

which leads to:

$$\boldsymbol{QT}(i+1) = \frac{QT_B - \boldsymbol{QT}(i)}{e^{-\lambda iT}} \cdot (1 - e^{-\lambda T})e^{-\lambda iT} + \boldsymbol{QT}(i) \tag{9.6}$$
$$= (QT_B - \boldsymbol{QT}(i)) \cdot (1 - e^{-\lambda T}) + \boldsymbol{QT}(i) \tag{9.7}$$
$$= QT_B(1 - e^{-\lambda T}) + e^{-\lambda T} \cdot \boldsymbol{QT}(i) \tag{9.8}$$

In this equation, a new $QT(i+1)$ depends on the old $QT(i)$. The first term in Equation 9.8 includes the value to which QT adapts, QT_B. As this value is dependent on RR, as every correction formula in any way assumes, the new QT interval is dependent on both, QT and RR.

An example using correction formula of Bazett follows:

$$QT_c = \frac{QT}{\sqrt{RR}} \tag{9.9}$$

$$\tag{9.10}$$

With the assumption that formula of Bazett is valid after full adaptation of QT:

$$QT_B = QT_c \cdot \sqrt{RR_B} \tag{9.11}$$

Equation 9.8 can be written as:

$$QT(i+1) = QT_c(1 - e^{-\lambda T})\sqrt{RR_B} + e^{-\lambda T} \cdot QT(i) \tag{9.12}$$

In this equation, the dependency of $QT(i+1)$ on $QT(i)$ and RR_B can be seen. This equation motivates using an autoregressive process (AR process) to describe the dynamics between QT and RR.

9.2 MVAR-Model Estimation

9.2.1 Time Series Preconditioning

ECG data have been processed, using methods described in Chapter 7. The time series exported out of BSAT do not represent a time series with equidistant data points. This is a problem, as the estimation of the model parameters is time dependent. Hence, a preprocessing is necessary to transform the time series of QT- and RR interval data into time series with equidistant data points [115].

The non equidistant time series RR and QT have been up-sampled to a time series with a sample frequency of 10 Hz using a linear interpolation in Matlab. 10 Hz is a compromise between having enough data points to get a well representation of the original time series (with only small error) and keeping the estimation data as small as possible. The interpolated time series is afterwards filtered by a mean value filter. The linear interpolation leads to discontinuities in the time series, which are banned by the filtering. The mean value filter has a window length of 5 samples. The resulting time series are named as $\widetilde{RR}$ and $\widetilde{QT}$, respectively.

9.2.2 AR Model

In statistical analysis and signal processing, an AR model is a random process, which is used to describe a physical appearance. It can be used to describe a coupling between input data or to predict new data. The output of the model is based on the previous inputs. Equation 9.13 shows the basic description of a first order AR model.

$$X(t) = c + \sum_{p=1}^{P} \alpha_p \cdot X(t-i) + \varepsilon(t) \tag{9.13}$$

where X_t is the t-th sample of the time series X. The value P represents the order of the model and α_p are the model parameters. The value c is a constant which can be set to zero, while ε_t represents the white noise of the process.

By incorporating multiple time series, a more complex MVAR model is required.

$$X_t = BZ_t + \varepsilon(t) \tag{9.14}$$

with:

$$X_t = \begin{bmatrix} x_1 \\ \vdots \\ x_n \end{bmatrix}, \qquad B = [A_1 \ldots A_p], \qquad Z = \begin{bmatrix} x(t-1) \\ \vdots \\ x(t-P) \end{bmatrix} \tag{9.15}$$

$$A_p = \begin{bmatrix} \alpha^p_{1,1} & \cdots & \alpha^p_{1,n} \\ \vdots & \ddots & \vdots \\ \alpha^p_{n,1} & \cdots & \alpha^p_{n,n} \end{bmatrix} \tag{9.16}$$

To describe QT/RR coupling, a first order two dimensional MVAR model is introduced. The constant c is omitted for simplicity in all further considerations. White noise, ε_t, is also neglected in this process. Equation 9.17 shows the basic mathematical description in vector notation.

$$\begin{bmatrix} \widetilde{RR} \\ \widetilde{QT} \end{bmatrix}_t = \begin{bmatrix} \alpha_{1,1} & \alpha_{1,2} \\ \alpha_{2,1} & \alpha_{2,2} \end{bmatrix} \cdot \begin{bmatrix} \widetilde{RR} \\ \widetilde{QT} \end{bmatrix}_{t-1} \tag{9.17}$$

The corresponding difference equation is given by:

$$\widetilde{\boldsymbol{RR}}(t) = \alpha_{1,1} \cdot \widetilde{\boldsymbol{RR}}(t-1) + \alpha_{1,2} \cdot \widetilde{\boldsymbol{QT}}(t-1) \tag{9.18}$$

$$\widetilde{\boldsymbol{QT}}(t) = \alpha_{2,1} \cdot \widetilde{\boldsymbol{RR}}(t-1) + \alpha_{2,2} \cdot \widetilde{\boldsymbol{QT}}(t-1) \tag{9.19}$$

In this model, four parameters, $\alpha_{1,1}$ to $\alpha_{2,2}$, have to be estimated. The parameters give information on the dependency between the time series. All coefficients α are found using maximum likelihood estimation [116].

9.2.3 Model Validation Using Artificial Time Series of QT and RR

To test the behaviour of the model estimation of time series from RR- and QT intervals, artificial time series have been generated. In the time series, RR is changed in a step and the adaptation of QT follows an exponential function according to Section 9.1.2.

If the model describes the coupling between both input time series, the exponential rate constant λ can be returned by the model parameters. Equation 9.20 shows the calculation of λ out of the coefficient $\alpha_{2,2}$.

$$\lambda = \frac{-ln(\alpha_{2,2})}{T} \tag{9.20}$$

Rate constant λ of several different artificial time series have been successfully reconstructed during periods with no change in $\widetilde{\boldsymbol{RR}}$. A change in $\widetilde{\boldsymbol{RR}}$ has led to a change in the model coefficients and a successful reconstruction was not possible. Nevertheless, the model describes the relationship between QT and RR successfully.

To check the quality of the estimated model, it has been used to predict $\widetilde{\boldsymbol{QT}}$ after a step in $\widetilde{\boldsymbol{RR}}$. Figure 9.2 shows the artificial time series of $\widetilde{\boldsymbol{RR}}$ and $\widetilde{\boldsymbol{QT}}$. The sequence is split into two sections. The time between heart beat 150 and 190 was used to estimate the model parameters. Then, the following QT-values have been predicted using the previously estimated model parameters. The blue curve in Figure 9.2(a) shows the prediction of the QT-values. Until the step in $\widetilde{\boldsymbol{RR}}$ at heart beat 200, the forecast follows $\widetilde{\boldsymbol{QT}}$ with only small error. After the step in $\widetilde{\boldsymbol{RR}}$, the forecast fails. The change in $\widetilde{\boldsymbol{RR}}$ does not lead to a change of the curve of predicted QT-values. The model describes the dynamics of the system, but does not incorporate external influence.

To incorporate changes, the system needs to be enhanced by an innovation matrix, describing the external changes:

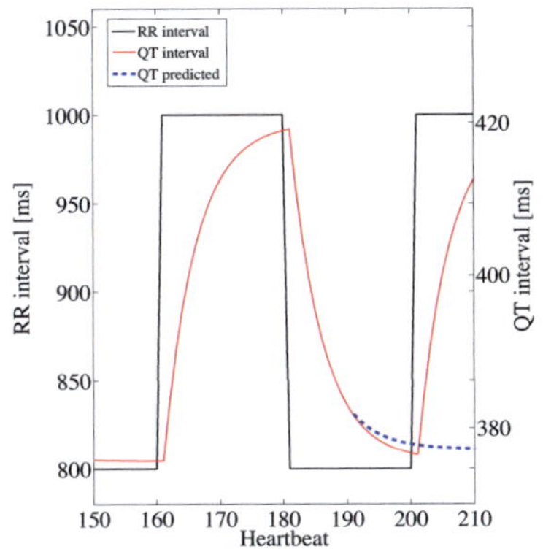

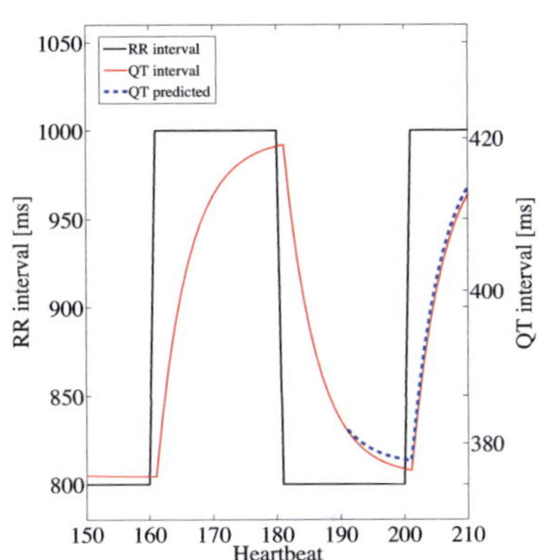

(a) Predicted time series without using innovations

(b) Predicted time series using innovations

Fig. 9.2. Artificial QT time series prediction with and without consideration of innovations.

$$\begin{bmatrix} \widetilde{RR} \\ \widetilde{QT} \end{bmatrix}_i = \begin{bmatrix} \alpha_{1,1} & \alpha_{1,2} \\ \alpha_{2,1} & \alpha_{2,2} \end{bmatrix} \cdot \begin{bmatrix} \widetilde{RR} \\ \widetilde{QT} \end{bmatrix}_{i-1} + \begin{bmatrix} I_{RR} \\ I_{QT} \end{bmatrix}_i \tag{9.21}$$

The changes in $\widetilde{QT}$ come from changes in $\widetilde{RR}$. Hence, the innovation process is only dependent on the time series of RR. The innovation time series of QT is set to zero.

$$\begin{bmatrix} I_{RR} \\ 0 \end{bmatrix} \tag{9.22}$$

Using this system with innovations led by $\widetilde{RR}$, a successful prediction after a step in $\widetilde{RR}$ is possible. Figure 9.2(b) shows the result.

The introduced model can describe the dynamics between two time series, which have been artificially generated and are coupled by an exponential function. Next, time series resulting from real ECG data have been analysed according to the coupling of QT and RR. The results are presented in the next section.

9.3 Heart Rate Dependency of Model Coefficients

To evaluate the QT/RR coupling in healthy subjects over a wide heart rate range, all 10 recordings of IBT-Exercise-study have been used to estimate model coefficients. A first order MVAR model was used. The window size has been 500 samples, while the shift between two successive estimations was 50 samples. A time lag of 20 samples has been used. Input data were time series $\widetilde{RR}$ and $\widetilde{QR}$, which have been preprocessed as described in Section 9.2.1 .

As the QT interval is adapted to the RR interval and not vice versa, analysis of the coefficients $\alpha_{1,1}$ and $\alpha_{2,1}$ is not reasonable. Hence, the focus is on the coefficients $\alpha_{2,1}$ and $\alpha_{2,2}$ in this section.

In Figure 9.3 and 9.4, the coefficients are presented depending on the corresponding mean RR interval length of the estimation window, for two different participants. It can be observed that $\alpha_{2,1}$ lies constantly at about zero for RR intervals above 700 ms, while $\alpha_{2,2}$ lies constantly at about one. This could be observed in all subjects of the study.

For RR interval lengths below 700 ms, the coefficients start to change. While $\alpha_{2,1}$ increases to values around one, $\alpha_{2,2}$ decreases to values near zero for high heart rates.

Hence, the QT interval of a heart beat i is strongly coupled to the previous QT interval of heart beat $i - 1$ for low heart rates (RR intervals above 700 ms). The influence of the last RR interval is small. For high heart rates, the QT interval is starting to get more and more dependent on the previous RR intervals. While the influence of RR increases during sporting activity, the influence of previous QT intervals decreases.

Changes in the parameters are coupled to the load of the heart. It can be assumed, that the relationship between QT and RR is changed during sporting activity. In Section 8.4, a change of the T wave was detected on the same data. Changes in the T wave are assumed to be related to a modified repolarisation. This change of repolarisation might also affect the QT/RR coupling, which can be observed by analysing the coefficients of the MVAR model.

9.4 Course of Disease Analysis Using MVAR Model Estimation

To analyse the course of disease of a myocarditis, eight ECG recordings of about 45 minutes duration from one subject have been analysed. The ECGs have been recorded at intervals of one week. The recordings are named A to H. The first recording (A) has been made at hospitalisation.

Again, the ECG signals have been processed by the methods described in Section 7. The resulting time series of QT- and RR intervals have been preprocessed as described in Section 9.2.1. A first order MVAR model with time lag of 20 samples has been used. The estimation window had a size of 500 sample points. The window was shifted after each estimation by 50 samples.

Figure 9.5 and 9.6 show the estimated model parameters $\alpha_{2,1}$ and $\alpha_{2,2}$[1] for all recordings in relation to the mean heart rate of the estimation window. A clear dependency between heart rates and the values of the coefficients is not observed. It can be seen that the variance of both coefficients is widely spread for all heart rates in the recordings A to F. While $\alpha_{2,1}$ is mostly concentrated between 0 and 0.5, $\alpha_{2,2}$ is spread between 0 and 1.

---

[1] As previously described, $\alpha_{2,1}$ and $\alpha_{2,1}$ describe the physiologically meaningful relations between QT and RR.

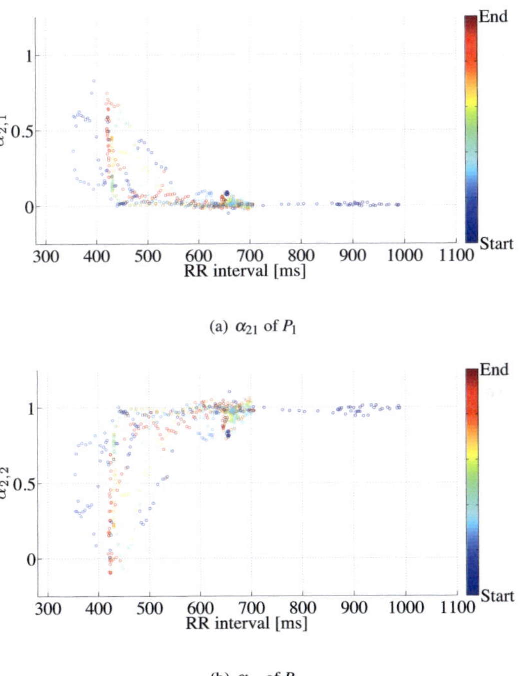

(a) α_{21} of P_1

(b) α_{22} of P_1

Fig. 9.3. Heart rate dependency of MVAR model parameters for participant P_1. Input data, RR and QT have been up-sampled to equidistant time series with sample frequency of f_S=10 Hz. Time lag of the model was set to 20 values (2 sec), window size was set to 500 values, shift between estimations was 50 values. Colour of the cycles represent the time during the recording.

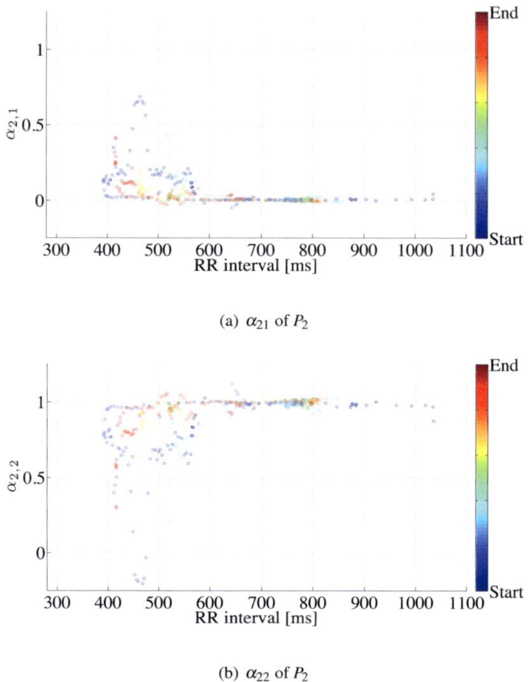

(a) α_{21} of P_2

(b) α_{22} of P_2

Fig. 9.4. Heart rate dependency of MVAR model parameters for participant P_2. Input data, RR and QT have been up-sampled to equidistant time series with sample frequency of f_S=10 Hz. Time lag of the model was set to 20 values (2 sec), window size was set to 500 values, shift between estimations was 50 values. Colour of the cycles represent the time during the recording.

In the last two recordings H and G, the variance of the parameters gets smaller. While $\alpha_{2,1}$ turns to 0, $\alpha_{2,2}$ turns to 1. A heart rate dependency is neither seen. To get a better impression of the median values and the variance, Figure 9.7 shows boxplots of all eight recordings. A decrease in the variance of the last two recordings is clearly visible for both parameters. The variance in recording F is not as small as in recording G and H, but smaller than in the previous recordings.

The median values of $\alpha_{2,1}$ seem to be decreased for the last three recordings, while the median values of $\alpha_{2,2}$ are clearly increased in recordings F to H.

The values of recording E differ clearly from all others. The highest variance for both parameters can be observed for recording E.

To analyse the coupling between QT and RR in different states of the course of the disease, the behaviour of the coefficients was analysed using the spectrum of the estimated coefficients.

Coefficient Spectrum

The time series of model coefficients $\alpha_{2,1}$ and $\alpha_{2,2}$ for each recording have been Fourier transformed and the absolute value of the Fourier transform has been observed. The FFT spectrum of the coefficients gives information on how fast the coefficients change. Fast changes of the model parameters lead to an increase of

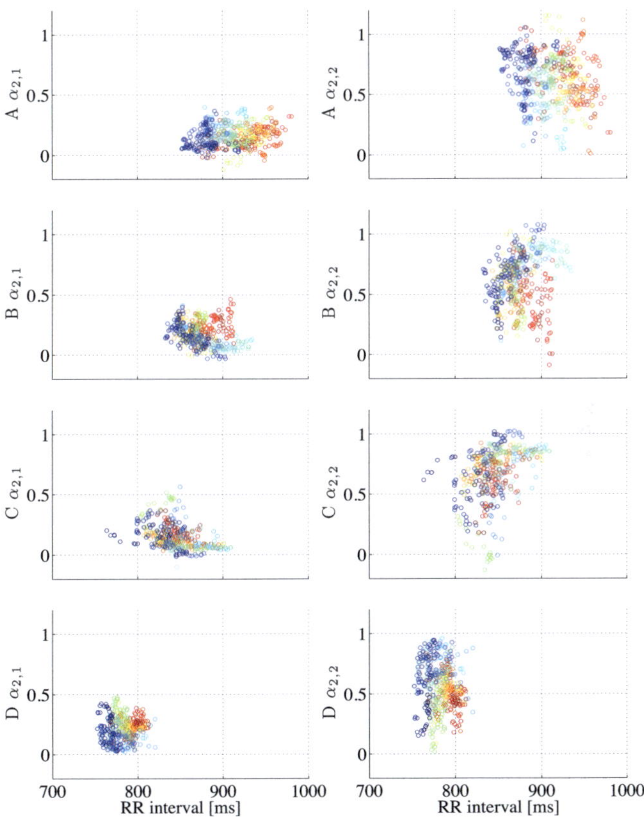

Fig. 9.5. Estimated MVAR model parameter over history of disease in dependence of heart rate (recording A-D).

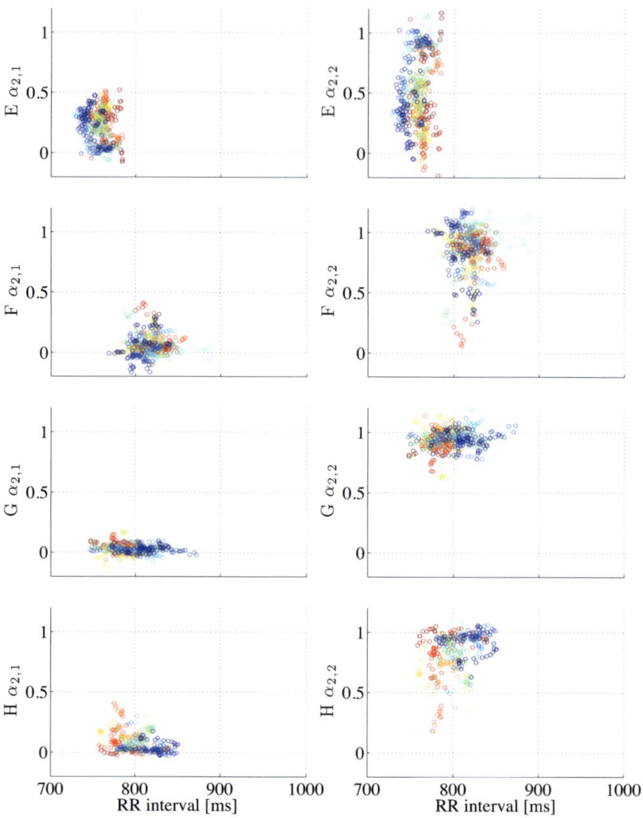

Fig. 9.6. Estimated MVAR model parameters over history of disease depending on the heart rate (recording *E-H*).

the values at the centre of the FFT spectrum, while slow changes lead to increased values at the beginning and end of the spectrum. Constant coefficients do not cause a decrease in the spectrum.

Figure 9.8 shows the result of all recordings and both parameters as an image graphic plot. In this plot, all areas are coloured corresponding to the matrix value. Low values of the spectrum are visualized in blue, while high values are visualized in red. It can be observed that the activity of the model parameters seems to be higher in the first recordings *A* to *F*. Orange, red and dark red areas are more often observed in these recordings. Recording *G* and *H* seem to have a lower activity.

Again, activity of both parameters is lowest in recording H. Recording E is the recording with highest activity of both parameters.

The activity of the parameters applies to the results taken from the boxplots. Recording F has a lot of outliers in the boxplot. These might be responsible for the high activity of both parameters.

Altogether, a clear difference in the QT/RR coupling between the first and the last two recordings can be observed. As the recordings have been made during the course of the disease, the healing of myocarditis might be related to the results. The investigations in Section 9.3 showed that model parameters of healthy subjects are more or less constant for heart rates below 86 bpm (700 ms RR interval length). The heart rate in all eight recordings (A to H) of the Myocarditis-study have been mostly under the limit of 86 bpm. Only results of recording G and H are very similar to those of the healthy subjects in IBT-Exercise-study, whereas recording G fits best those results. Recording A to F do not fit to the results of Section 9.3. As the recording show a history of the disease the more stable coefficients in recording G and H might be related to a recovery from disease.

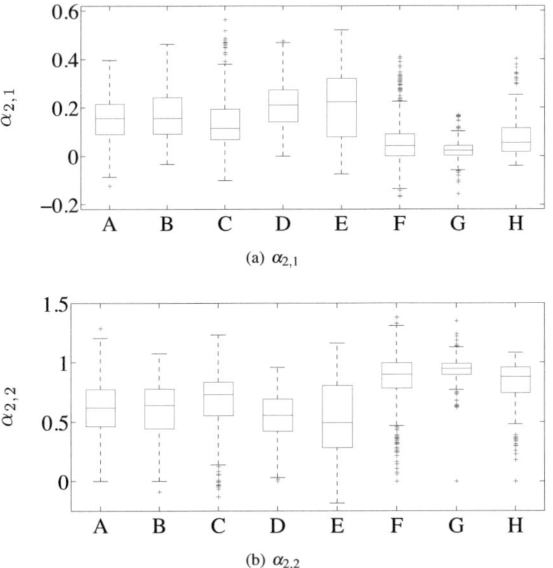

(a) $\alpha_{2,1}$

(b) $\alpha_{2,2}$

Fig. 9.7. Boxplot of MVAR parameter estimation over history of disease.

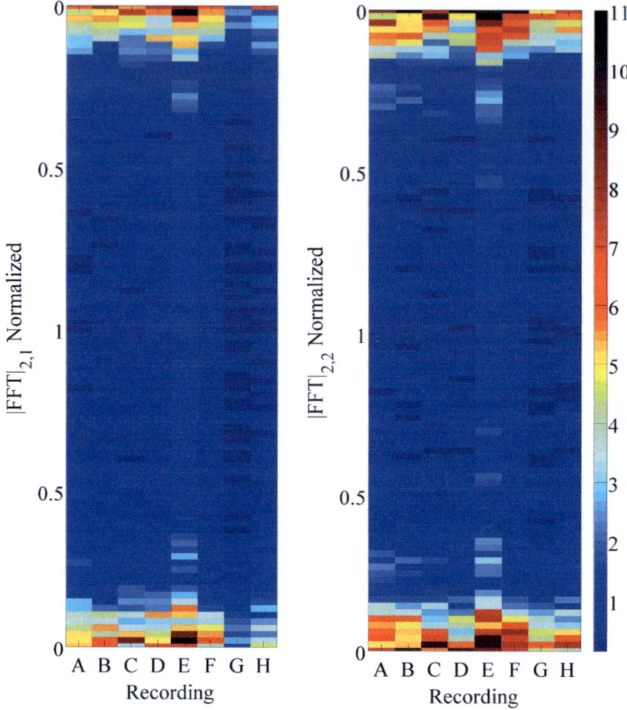

Fig. 9.8. FFT spectrum of estimated MVAR model coefficients during the course of myocarditis.

9.5 Discussion

In this chapter, the focus was on the analysis of the QT/RR relationship by an MVAR model. The analysis of the ventricular repolarisation in the surface ECG is often related to the QT interval and the coupling between QT and RR. The method described in this chapter provides a way of analysing the coupling between QT and RR, to get new information on the repolarisation process. The information was obtained by analysing model coefficients of a MVAR model. Changes in the repolarisation process should affect the QT/RR coupling and this results in a change of the model parameters.

The ECG data of the IBT-Exercise-study has been analysed using an MVAR model. It turned out that heavy physical load leads to a change in the model parameters, while the coefficients stay more or less constant at rest. The relationship between QT and RR seems to be dynamic and is affected by heavy sporting activity. Changes in the repolarisation have also been identified in Section 8.4 by analysing the morphology of the T wave. Although the number of subjects was small in this investigation, a trend was seen. Further investigations using a higher number of subjects with additional repeated recording of each subject is recommended to confirm the results.

The heart is assumed to be in a stationary state during one estimation. As the estimation window was 500 samples (50 seconds), the stationarity might be disturbed for very high heart rates and thus the parameters change so significantly. However, the steady state for heart rates above 85 bpm (700 ms RR interval length and more) should be given.

The settings of the model should be investigated in further research. In [10] several investigations with different settings have been made on the same dataset. It turned out that the results are dependent on the length of the analysing window, the lag and the order of the model.

The course of a disease was analysed using ECG data of one patient suffering from myocarditis. ECG recordings of eight weeks have been analysed. Modifications in the model parameters have been observed between the first weeks after hospitalisation and the seventh and eighth week. The model coefficients changed in variance, activity and median values. However, a clear evidence of describing the healthy state of the subject cannot be made, as the number of one analysed subject is by far not enough. For this subject, good results have been obtained. Thus, a study with more patients should be carried out in near future to confirm these results.

The way of getting new information out of ECG time series using MVAR models is promising. The returned model coefficients represent subject specific parameters of the state of the heart, as Malik claims in [117]. Using other time series, as e.g. related to the T wave morphology, could help to improve the detection of repolarisation changes. Therefore, analysing estimated model parameters of MVAR models is recommended for the investigation of heart diseases.

10

Influence of Ectopic Beats on the Repolarisation

In this chapter an investigation on the influence of ectopic beats on the repolarisation sequence of the ventricles is presented. ECG recordings from the PVC-study and additionally one continuous 10 day ECG recording were analysed. The recordings with different numbers of PVC and PAC were delineated and the beats classified as described in Chapter 7. A Heart Rate Turbulence analysis (HRT) has been performed on all PVCs fulfilling the rules of the HRT as they are defined in [118]. To detect the influence of the PVC on the repolarisation sequence, the T wave has been analysed regarding morphology changes. Descriptors introduced in Section 7.5.1.2 have been used to detect these T wave changes and thereby the underlying modifications of the ventricular repolarisation.

10.1 Heart Rate Turbulence

The heart rate turbulence analysis was introduced by Schmidt et. al [119]. In this method the fluctuation of the sinus rhythm after a PVC is described by two parameters. The absence of this fluctuation seems to be a risk predictor for sudden cardiac death after a myocardial infarction. In several studies the combination of both parameters turned out to be the strongest univariate risk predictor [118]. The HRT is independent from other known post-infarction risk stratifiers [119].

HRT-Parameters

The first parameter of the HRT is named *'Turbulence Onset'* (TO). It describes the difference between the mean of the first two RR intervals after the Compensatory Pause (CP) that follows a PVC and the mean of the last two RR intervals before the Coupling Interval (CI) of a PVC. The value is expressed as percentage of the difference to the post PVC mean RR intervals.

$$TO = \frac{(RR(i+1) + RR(i+2)) - (RR(i-3) + RR(i-2))}{(RR(i-3) + RR(i-2))} \cdot 100\% \qquad (10.1)$$

in this equation $RR(i)$ represents the compensatory pause and $RR(i-1)$ the coupling interval.

The second HRT parameter is named *'Turbulence Slope'* (TS). It describes the maximum positive slope of five subsequent RR intervals within the first 20 RR intervals that follow a PVC. The value of TS is expressed in milliseconds per RR interval. The calculation is done by a regression of five RR intervals. Figure 10.1 illustrates the HRT parameters TO and TS.

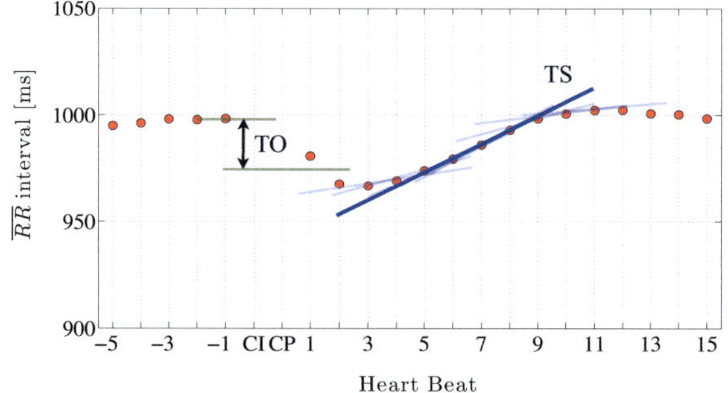

Fig. 10.1. Illustration of the Heart rate turbulence parameter. Mean values come from 590 PVCs in a 10 day Holter ECG recording. Coupling interval labelled as CI, compensatory pause labelled as CP.

PVC Filter

The computation of the HRT parameter can only deliver usable results, if all involved PVCs have been correctly classified. If artefacts or large amplitude T waves are considered due to misclassification, the method fails. Therefore, rules have been introduced which must be fulfilled to take a PVC into account for the HRT analysis. First of all, five normal beats before the coupling interval of a PVC and 20 normal beats after the compensatory pause are claimed. All beats in this interval have to fulfil the following rules:

- $RR_n > 300\,ms$
- $RR_n < 2000\,ms$
- $|RR_n - RR_{n-1}| > 200\,ms$
- $RR_n \leq 0.8 \cdot \overline{RR}_n$, or $RR_n \geq 1.2 \cdot \overline{RR}_n$, with: $\overline{RR}_n = \frac{1}{5} \cdot \sum_{j=1}^{5} RR_{n-j}$.

Otherwise the PVC cannot be considered in the HRT analysis. Additionally the PVC itself needs to fulfil the following rules:

- Coupling interval needs to be smaller than 80% of the corresponding $\overline{RR}$
- Compensatory pause must be at least 20% larger than the corresponding $\overline{RR}$

Further information on the computation of the HRT parameters can be found in [120, 121].

Clinical Interpretation of the HRT

For the interpretation of TO and TS, both values can be used as separate clinical variables or in combination [120]. The values of TO is interpreted in most clinical studies as normal for $TO < 0\%$. Accordingly TS is interpreted as normal for $TS < 2.5 \frac{ms}{RR\,interval}$. Three classes are ordinarily used for classification of both variables in common:

- **HRT category 0**: TO and TS are normal[1].
- **HRT category 1**: One of the variables TO or TS is abnormal.
- **HRT category 2**: Both variables TO and TS are abnormal.

HRT Results

The HRT of several subjects in the PVC-study have been calculated. Besides the PVC study, a 10 day Holter ECG recording of one subject has been analysed. The recording is named H-10D. Figure 10.2 shows the HRT of the H-10D recording. A classical HRT slope can be seen.

A validation of the HRT results was not possible because of the lack of patient information in PVC-study. However, the computation of the HRT parameter was successful done for all patients with more than 15 suitable PVCs. Table 10.1 shows the resulting HRT parameters TO and TS. These parameters have been calculated twice, once separated for each time series in the recording with a subsequent calculation of the mean value of all resultant parameters of the recording and once, by calculating an average RR interval series of the recording and subsequent calculation of the HRT parameters on this averaged time series. The results of TO are quite equal for both methods, while the results for TS were different. In [120] similar results have been reported and it is recommended to calculate an average RR time series to calculate TS.

---

[1] In case of too few suitable PVCs in the recording, but normal sinusrhythem in the rest of the signal, the recording is also category 0.

Table 10.1. Results of HRT analysis. The values were calculated by the mean of HRT values $(\overline{TO},\overline{TS})$ and by calculating a mean RR time series and consequential calculation of the HRT parameters $(TO_{\overline{RR}},TS_{\overline{RR}})$. The categories are related on $TO_{\overline{RR}}$ and $TS_{\overline{RR}}$

Patient	No. PVC	$\overline{TO}$	$\overline{TS}$	$TO_{\overline{RR}}$	$TS_{\overline{RR}}$	Category
P-1	80	-1,17	5,48	-1,23	2,58	0
P-21	25	0,92	10,93	0,91	5,42	1
P-22	26	-4,76	11,54	-4,93	5,52	0
P-23	35	0,09	6,94	-0,02	3,46	0
P-24	21	-0,59	3,64	-0,71	0,89	1
P-29	15	1,10	2,36	1,11	1,12	2
P-49	33	-0,18	7,37	-0,19	3,46	0
P-50	42	-1,58	10,27	-1,65	5,47	0
H-10D	590	-2,35	10,83	-2,38	6,23	0

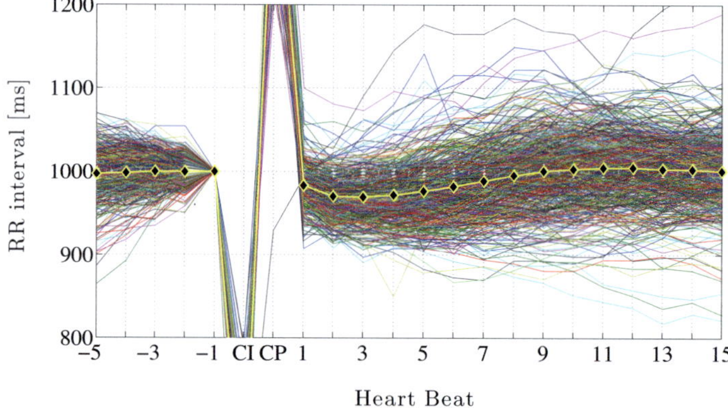

Fig. 10.2. Heart rate turbulence in a 10 day Holter ECG recording. The values were normalized to 1000 ms for the last normal heart beat before the PVCs. A total of 590 PVCs have been considered. The yellow line shows the resulting average HRT slope.

The results of the PVC-study have to be seen in the context of a short recording time of only one hour. In order to get a meaningful number of suitable PVCs, normally 24h-Holter ECG recordings are used for HRT analyses.

The HRT considers the variation of the RR intervals following a PVC. The change of the RR intervals due to the PVC is assumed to be a reaction of the autonomic nerve system. It is supposed that the blood pressure plays a major role. However, the influence of a PVC to the de- and repolarisation sequences of subsequent heart beats is not yet considered, but might also be an interesting maker.

10.2 Morphological Descriptors of the T-Wave

To analyse potential repolarisation changes of heart beats following a PVC, all morphological descriptors introduced in Section 7.5.1.2 have been computed for 5 heart beats before the coupling interval and 15 heart beats that follow the compensatory pause. For reasons of clarity 6 morphology based descriptors have been chosen to be presented here.

Results

Figure 10.3 shows the results of subject P1 of the PVC-study, while Figure 10.4 shows the result for 150 consecutive PVCs out of the 10 day recording H-10D. In both Figures the curve of the RR intervals following the PVC has a typical shape. The QT interval in Figure 10.3 is only changed during the compensatory pause, which is related to the different shape of the T wave in this interval. The first two beats after the compensatory pause might be slightly affected. In Figure 10.4 an increase of the QT interval with a decrease to the old values might be sensed. However, a clear trend cannot be deduced for the QT interval.

In addition to the RR- and QT interval, morphology depicting descriptors are shown in Figure 10.3 and 10.4. It can be seen that the T wave one beat after the PVC is clearly changed. Nearly all parameters show a modification compared to the values before the PVC and some beats later. Later following beats show only a small difference compared to the baseline. The return to the baseline values is very fast in this subject. The amplitude of the T wave seems to take 3 to 4 heart beats to get stable again. Altogether the influence of the PVC to T waves of subsequent heart beats is high for the first beat and decreases very fast to zero within the next 2 or 3 heart beats.

In Figure 10.4 the influence of the PVC on the repolarisation of the subsequent heart beats seems to be enlarged. A PVC modifies the T waves describing parameters for up to 7 heart beats. Interestingly the return to the baseline values can be clearly seen as a characteristic slope.

During the analysis of all morphology based descriptors the time dependent values δ_I to δ_{III} showed only little modifications after a PVC. The values of the area based descriptors Λ_I to Λ_{III} showed similar behaviour as the amplitude value ν. This is

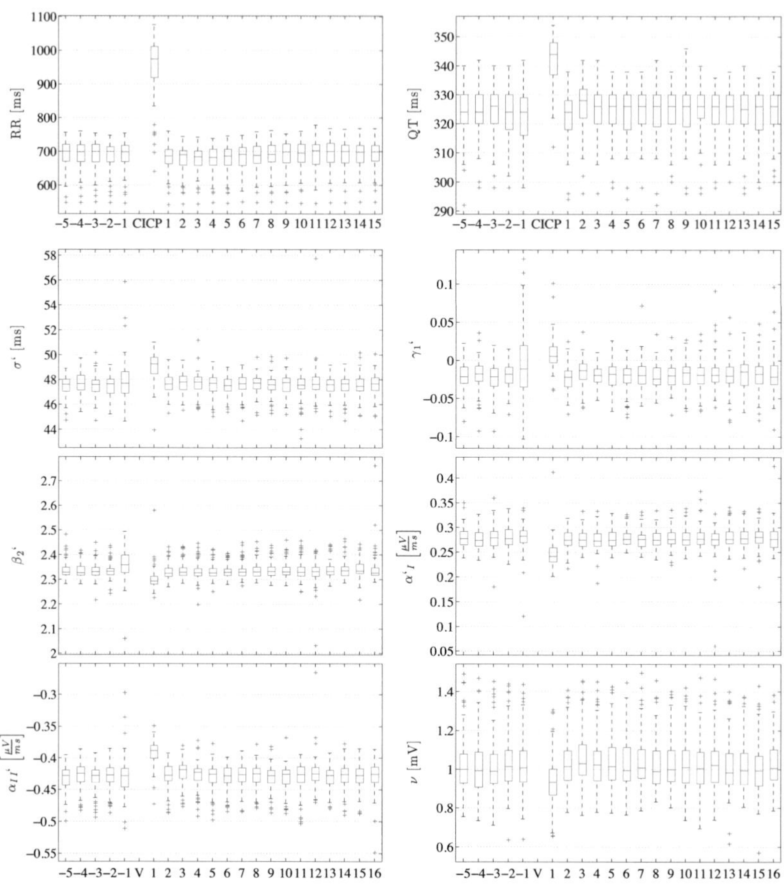

Fig. 10.3. Behaviour of RR-, QT interval and T wave morphology depicting descriptor after the appearance of a PVC in subject 1. Values of the PVC have been neglected. 42 PVCs have been considered.

not surprising, as the amplitude is changed and this significantly influences the area based values. Figure 10.5 shows a 3D plot of mean T waves for the analysed heart beats around the PVCs. The mean T waves were computed from the corresponding waves of the 590 PVCs in the 10 day recording.

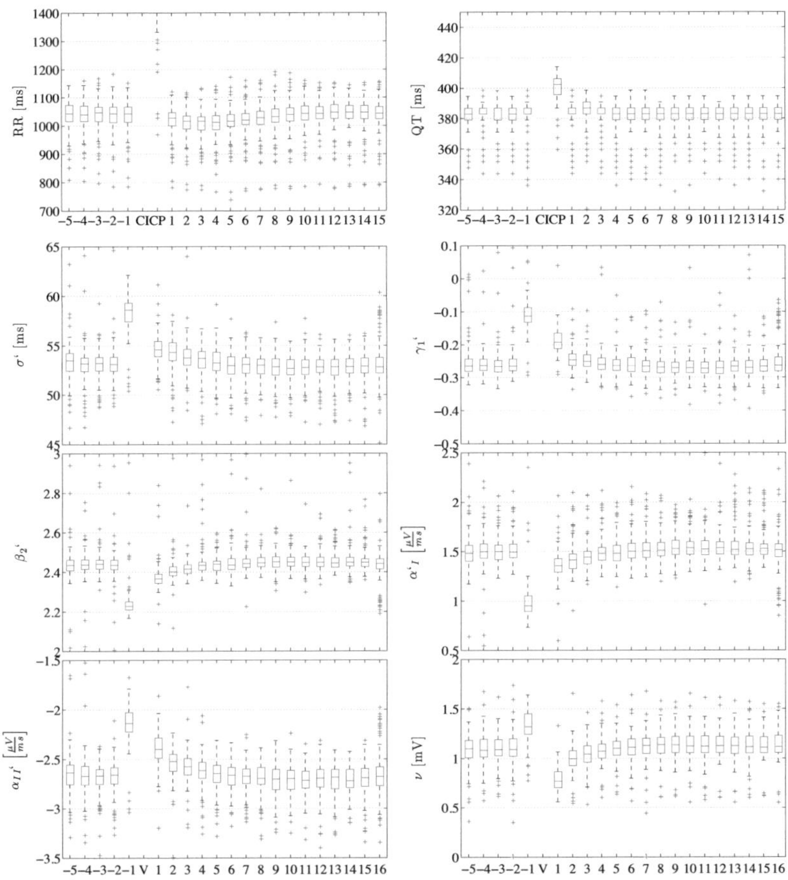

Fig. 10.4. Behaviour of RR-, QT interval and T wave morphology depicting descriptor after the appearance of a PVC in subject 2. The values of the PVC have been neglected. 150 PVCs have been considered.

Influence of the Length of the Coupling Interval and the Length of the Compensatory Pause

To investigate, whether the length of the coupling interval or the length of the compensatory pause influences the results, the PVCs of the 10 day recording H-10D have been classified accordingly. Nine classes have been defined out of the combinations of: three classes divided from the coupling interval and three classes

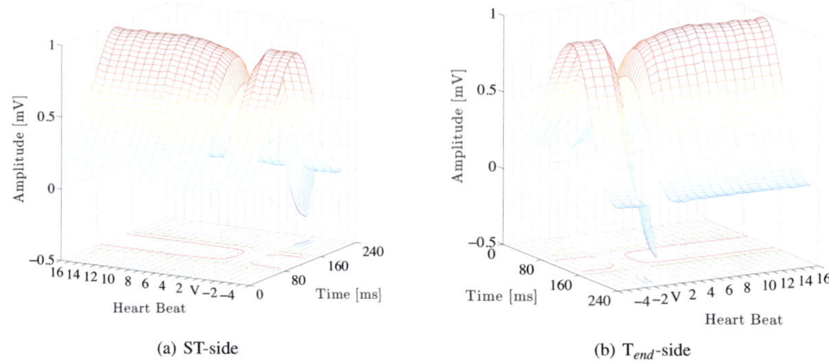

(a) ST-side   (b) T_{end}-side

Fig. 10.5. 3D plot of the mean T wave of an assembly around all PVCs in a 10 day Holter ECG recording. ST-side of the T wave on the left, T_{end}-side on the right. Mean PVC-wave has been blanked. The last T wave before the PVC is decreased at the end. This is due to the influence of the PVC directly behind the T wave of the last heart beat before the PVC.

divided from compensatory pause. Figure 10.6 shows the results for the amplitude in the recording H-10D. In X-direction, a separation on the Coupling Interval (CI) is made, while in Y direction a separation on the Compensatory Pause (CP) is made.

As already mentioned, the amplitude of the T wave is decreased during the PVC and subsequent 5 to 7 successive heart beats. This can be seen in nearly all axes of the Figure 10.6. A significant difference between the classes is not noticeable. Additionally, it has to be considered that the number of the PVCs is very different among the axes. Observed variance in the parameters might be related to this fact. However, a clear result is not observed.

All of the T wave describing parameters that have been introduced in Section 7.5.1.2 have been investigated in the same manner as the T wave amplitude. A clear trend was not discovered in any of the parameters.

10.3 Discussion

In this chapter the HRT analysis was made for the data of the PVC-study. It has been confirmed that RR intervals following a PVC react in a typical manner. The HRT parameters have been calculated for all recordings in the PVC-study which had at least 15 PVCs fulfilling the rules of the HRT analysis. Additionally the HRT analysis was used for a 10 day Holter ECG recording. A validation of the HRT results was not possible, because of the missing medical history of the subjects from which the ECG data was available.

To analyse the influence of a PVC on the repolarisation process of the human heart, T wave describing parameters have been analysed for T waves that followed a PVC. It was shown that a PVC can change the T wave for up to 7 heart beats. Analysing the amplitude of the T wave seems to be most promising. Nevertheless, other T wave morphology depicting parameters have been affected by the PVC as well. However, it turned out that the effects of a PVC on subsequent T waves is different between the subjects. For instance the number of affected T waves was different between different recordings. It should be kept in mind that also the number of ectopic beats was different. Although the 10 day Holter ECG involves most PVCs of all recordings with 590 PVCs, the mean difference between two successive PVCs is larger than in most other recordings. Unfortunately, the absence of the medical history makes the interpretation of the results difficult. However, the analysis of PVC induced repolarisation changes was possible and might help to further understand the consequences of PVC with respect to cardiac health.

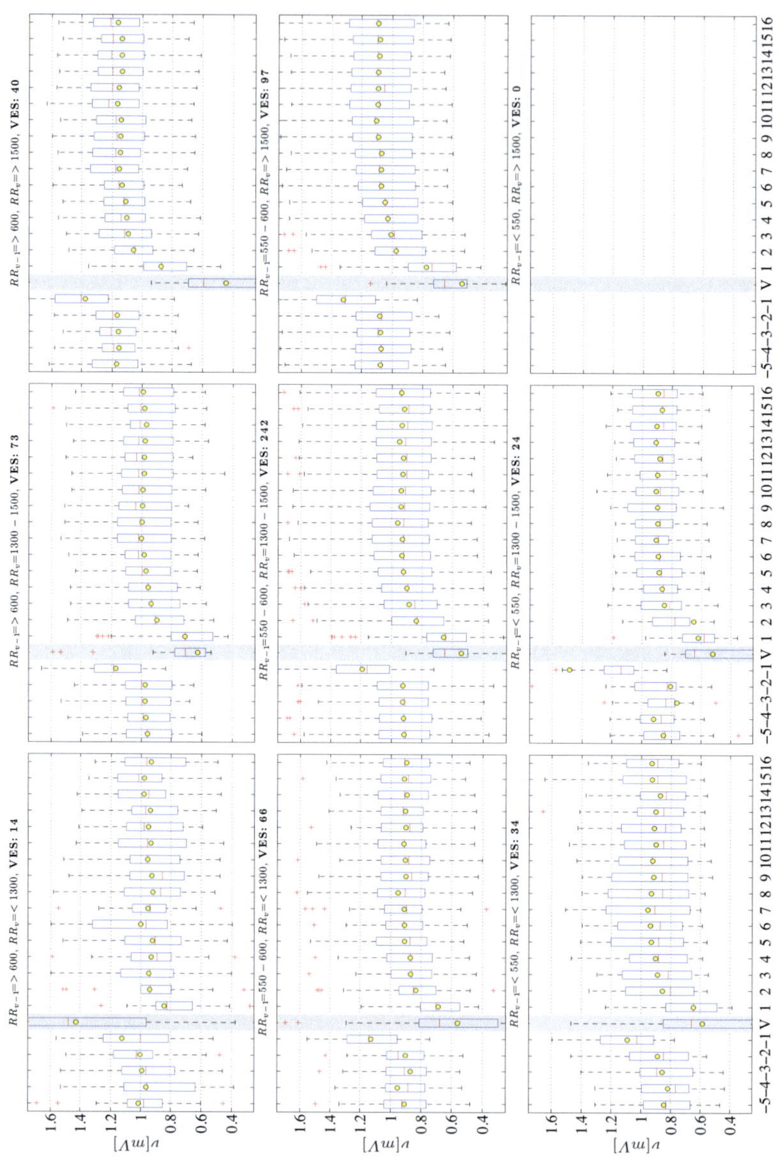

Fig. 10.6. Behaviour of the T wave amplitude after a PVC, classified by the length of both, coupling interval and compensatory pause. The yellow dots represent the parameter values calculated from an average T wave out of the T waves in the corresponding interval. A high correlation between the median values in the boxplot and the parameter values from the averaged waves can be observed.

Part IV

Closing Matters

11

Conclusions and Outlook

11.1 Conclusions

The presented thesis focuses on the ECG based investigation of the ventricular repolarisation in the human heart. Therefore, new methods in the field of ECG signal processing are introduced. A wavelet based delineation of the QRS complex was improved by a signal specific threshold optimisation routine to reliably detect all QRS complexes including QRS complexes of ectopic beats. The delineation of the T wave boundaries was introduced by a newly developed correlation based method. A patient specific T wave template, generated during the delineation process, makes the precise and reliable delineation of the T waves possible. The high accuracy delineation of the QRS- and T wave boundaries is the basis for the subsequent analysis of the ventricular repolarisation.

Beside the delineation of the ECG waveforms, a beat classification algorithm based on a support vector machine was introduced. The detection of ectopic beats, separated into premature atrial contraction and premature ventricular contraction was possible in two independent studies. The correct rate of the beat classification for multi lead delineation was 98.39% (SE 86.38%, PPV 77.25%). Based on the beat classification, a heart rate turbulence analysis (HRT) was performed on several ECG recordings.

The delineation of the ECG waveform and the beat classification have been used to generate time series. The time series can be divided into two classes. The first group comprises time based values, as e.g. RR- and QT interval, the second one comprises T wave morphology depicting parameters as e.g. T wave amplitude, skewness or kurtosis.

The whole ECG signal processing has been embedded into a software toolbox named *BSAT*. Using this toolbox, the introduced signal processing was performed to analyse the ventricular repolarisation.

In a first step, drug induced QT prolongation of some milliseconds could be detected in different drug safety studies for different compounds. Next, the analysis of the QT interval, claimed by the FDA as major risk stratifier in drug approval, has been extended by a detailed investigation of the morphology of the T wave. It turned out that both analysed compounds significantly changed several morphology depicting parameters in both studies. The significance has been tested by three statistical tests. Additionally, the blood plasma concentration, available in one study, showed good time based correlation to the change of most parameters.

To investigate the heart rate dependency of the morphology depicting parameters, an ECG exercise test study has been recorded. The study incorporates 10 healthy subjects. The recordings have been made using the ECG recorder *BSAT-1012* which has been developed and introduced especially for this purpose. It turned out that the T wave stays comparably stable for heart rates below about 85 bpm. Under heavy load in the exercise test, the T wave morphology was significantly altered, which could be observed as a change of the descriptors. However, the descriptors showed to remain stable at rest, which is a precondition for the analysis of drug safety studies.

Using the beat classification, an investigation of the modification of the ventricular repolarisation of heart beats subsequent to a premature ventricular contraction was performed. Although the number of analysed subjects was small and the medical history was missing, the results are promising. A change within up to 7 subsequent heart beats could be observed in one subject.

In another project of this thesis, the QT/RR coupling was in focus. The model parameters of a multivariate autoregressive model have been estimated and analysed. It turned out that under rest, the model parameters remained stable, while heavy load, due to sporting activity, led to a modification of the parameters. It seemed that the coupling was changed due to the loading and the high heart rate. Additionally, a subject suffering from myocarditis was analysed over the time of recovery, using the same MVAR model. A stabilisation of the estimated model parameters during recovery has been observed. Although only one patient was investigated, the results seemed to be promising.

11.2 Outlook

The analysis of the ventricular repolarisation using a set of parameters depicting the morphology of the T wave is recommended for drug safety studies by the author. The presented parameters seem to reliably describe different changes in the T wave morphology. A more detailed analysis of the changes of the ST-segment might improve the methods and provide additional information.

Using computer models of the heart, as they are presented in e.g. [19, 104], to simulate drug induced repolarisation sequences might help to better understand the underlying effects causing the observed T wave modification.

T wave morphology based descriptors could be used to investigate the ventricular repolarisation under the influence of heart diseases or psychological stress. This might eventually provide new diagnostic information. However, these studies require long term ECG data with detailed medical history of the investigated subjects to provide reliable results.

The detection of PAC and PVC in Holter ECG signals provides the opportunity to investigate effects of these beats on the excitation system of the heart. The meaning of the length of the coupling interval and the length of the compensatory pause should be further analysed, even though the results presented in this thesis showed no correlation. However, the number of analysed records was too small and the missing medical history limits the significance. Hence, an analysis using a clinical study with detailed medical history should be carried out in the near future. Again, computer models might help to better understand the underlying processes and should be considered in further projects.

The presented methods for delineating and analysing the T wave should be adapted to the P wave. This might help to diagnose atrial diseases and additionally give new information, which can help to improve the beat classification.

The estimation and analysis of MVAR model parameters, as exemplary shown in this thesis for one subject suffering from myocarditis, might help to analyse the condition of the heart and provide relevant diagnostic information in case of cardiac diseases. Extending the MVAR model to additional time series, as e.g. some of the introduced morphology based descriptors, will increase the model complexity, but might also provide new information on changes of the ventricular repolarisation sequence caused by a disease. This information might also be diagnostically relevant. Summing up, information resulting from estimated MVAR model parameters might help to dose medication in the clinical routine and get new information on the health condition during the recovery of a cardiac disease. Again, special Holter ECG data is necessary to prove the power of this new method and its clinical relevance.

It can be concluded that the methods introduced in this thesis should be tested on a higher number of clinically relevant ECG data. T wave morphology based descriptors should be used in clinical routine in the future.

A

Appendix

A.1 Results of T_{end} Delineation in MIT BIH QT Database

Table A.1. Results from the MIT BIH QT Database for the detection of T_{end}. Comparison between experts manual annotations against the fully automatic delineation of BSAT. The allowed maximum error was 100 ms.

Name	Beats	D_{Rate} [%]	$\overline{M}$ [ms]	$\overline{STD}$ [ms]	Lead	Name	Beats	D_{Rate} [%]	$\overline{M}$ [ms]	$\overline{STD}$ [ms]	Lead
sel100	30	100	-15.47	17.63	1	sel47	30	96.67	-24	16.84	1
sel102	63	100	-32.32	16.4	2	sel48	30	100	-72.67	12.65	2
sel103	30	100	-20.53	16.7	1	sel49	30	100	-16.13	13.95	1
sel104	45	77.78	9.94	36.07	2	sel50	31	96.77	19.2	16.42	1
sel114	49	100	-23.02	23.01	1	sel51	30	76.67	-68	16.14	1
sel116	50	100	-18.64	17.86	1	sel52	30	100	6.67	7.81	1
sel117	30	100	-23.6	15.09	1	sel803	30	100	-12.27	12.43	1
sel123	30	100	-14.67	10.04	1	sel808	30	100	17.87	19.22	1
sel16265	30	100	-16.67	16.04	2	sel811	30	100	-10.93	12.82	1
sel16272	30	100	0.93	10.7	1	sel840	70	100	-2.34	16.02	1
sel16273	30	100	1.73	10.28	1	sel871	70	100	-10.23	18.98	2
sel16420	30	100	4	10.18	1	sel872	30	100	-7.73	11.11	1
sel16483	30	100	-10.4	8.76	1	sel873	33	100	7.15	16.61	2
sel16539	30	100	-1.6	12.41	1	sele0104	30	73.33	-62.91	27.11	2
sel16773	30	100	-3.33	11.17	1	sele0106	30	100	-47.87	15.53	1
sel16786	30	100	-4.67	10.92	1	sele0107	34	100	-44.71	15.83	1
sel16795	30	100	-14.4	12.23	1	sele0110	30	96.67	-8.97	21.8	2
sel17152	30	100	-9.47	16.49	1	sele0111	30	100	-63.87	10.16	1
sel17453	30	100	2.27	20.2	1	sele0112	50	100	-21.76	9.5	2
sel213	60	98.33	-15.46	21.47	2	sele0114	30	100	-0.8	9.54	2
sel221	29	93.1	-20.15	19.8	2	sele0116	30	100	-25.07	23.16	1
sel223	31	70.97	-73.27	16.16	2	sele0121	30	100	-3.47	8.25	1
sel230	50	100	-4.08	17.38	2	sele0122	30	100	-1.33	5.79	1
sel231	50	100	-6.4	17.76	1	sele0124	50	100	-10.64	21.22	1
sel233	30	100	-8	13.78	1	sele0126	30	100	0.67	6.57	1
sel301	30	6.67	-96	5.66	2	sele0129	30	100	-28.67	11.56	2
sel302	30	100	-9.6	9.07	1	sele0133	30	100	-22.13	9.44	2

Name	Beats	D_{Rate} [%]	$\overline{M}$ [ms]	$\overline{STD}$ [ms]	Lead	Name	Beats	D_{Rate} [%]	$\overline{M}$ [ms]	$\overline{STD}$ [ms]	Lead
sel306	36	100	1.89	10.88	1	sele0136	30	96.67	-56.97	34.44	1
sel307	30	96.67	-5.24	17.07	1	sele0166	36	100	-31.33	17.9	1
sel308	28	96.43	-13.63	16.69	2	sele0170	30	100	-26.67	16.05	1
sel30	30	100	-18.67	12.35	1	sele0203	30	100	1.73	8.12	1
sel310	30	100	-7.87	8.63	1	sele0210	30	100	-12.93	13.23	1
sel31	30	80	39.17	47.34	2	sele0211	30	100	-4.8	10.73	1
sel32	30	100	-6.4	14.95	1	sele0303	30	0	∼	∼	1
sel33	30	96.67	-12.55	42.57	2	sele0405	30	100	-11.87	8.82	1
sel34	30	100	-11.87	10.63	1	sele0406	31	100	-17.16	12.57	1
sel36	31	0	∼	∼	1	sele0409	30	100	-28.53	9.55	1
sel38	15	6.67	96	∼	2	sele0411	30	100	-15.47	10.9	1
sel39	30	66.67	54.4	24.47	2	sele0509	30	100	-5.87	9.2	1
sel40	30	100	-1.6	17.15	1	sele0603	30	100	-10.4	10.32	1
sel41	30	13.33	-44	16.97	2	sele0604	30	100	2.27	9.08	1
sel42	30	100	-23.07	21.72	2	sele0606	30	100	-4.4	14.49	1
sel43	30	100	-1.87	8.65	2	sele0607	30	100	-30.93	10.02	1
sel44	22	54.55	-6.33	50.03	1	sele0609	30	60	-80	16.29	2
sel45	30	50	40.53	46.69	2	sele0612	30	100	-12.93	16.53	1
sel46	27	85.19	-66.43	23.13	1	sele0704	30	100	-0.93	9.67	1

A.2 Drug induced Repolarisation Change

A.2.1 THEW tQT$_I$-Study Moxifloxacin

Moxifloxacin induced changes of T wave morphology depicting descriptors of 34 subjects of *THEW tQT$_I$*-study. Interval length is 10 minutes. Only the first interval of every hour is shown. Time of dose is marked as D.

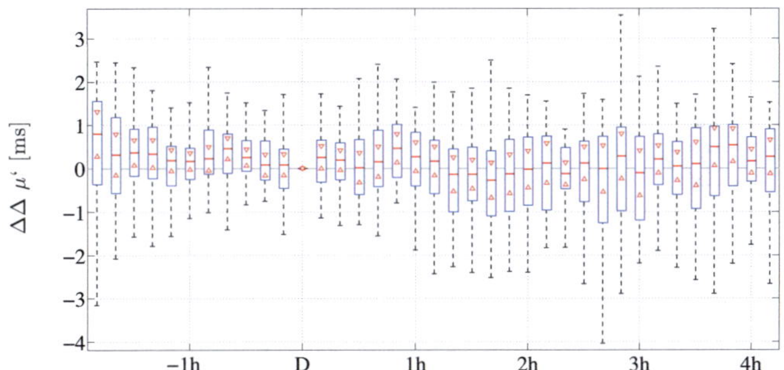

Fig. A.1. Boxplot of expected value, *THEW tQT$_I$*-study, Moxifloxacin

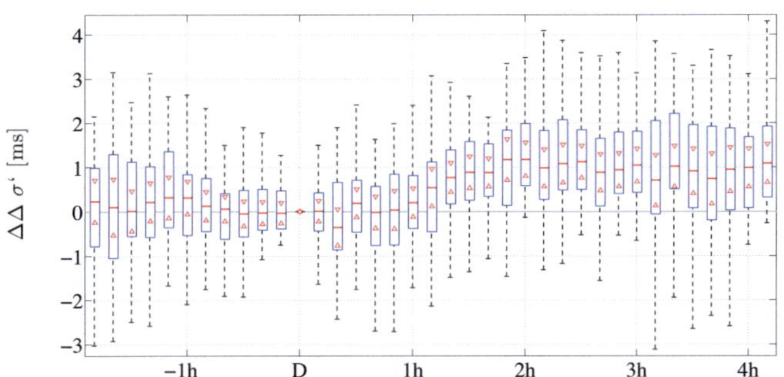

Fig. A.2. Standard Deviation, *THEW tQT$_I$*-study, Moxifloxacin

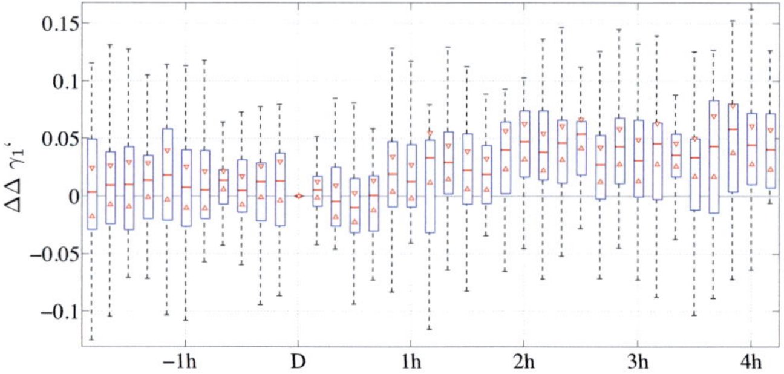

Fig. A.3. Boxplot of skewness, *THEW tQT_I*-study, Moxifloxacin

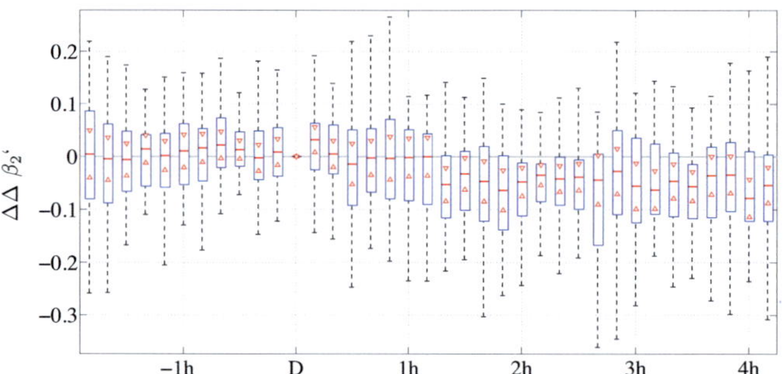

Fig. A.4. Boxplot of kurtosis, *THEW tQT_I*-study, Moxifloxacin

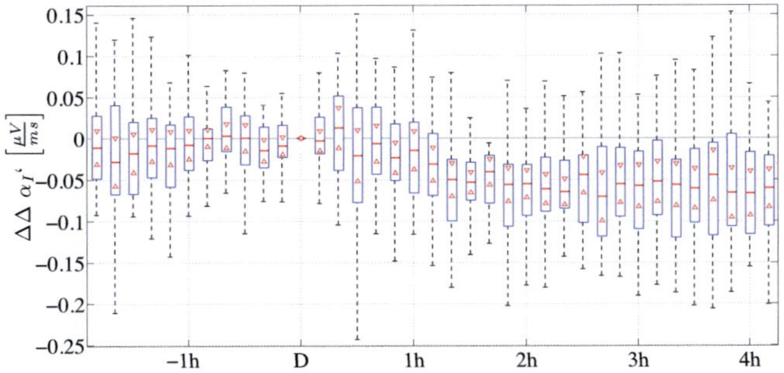

Fig. A.5. Boxplot of slope onset, *THEW tQT$_l$*-study, Moxifloxacin

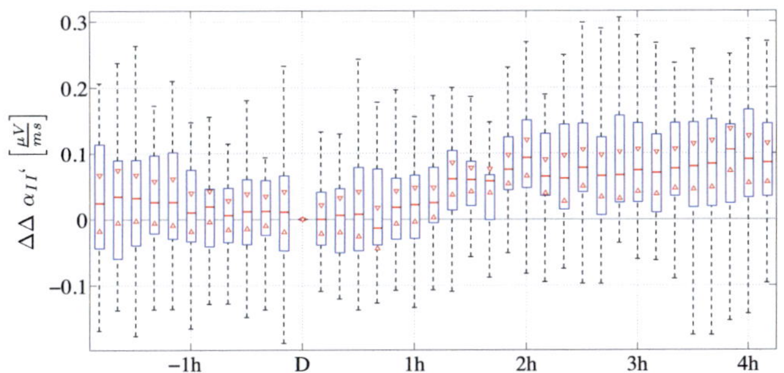

Fig. A.6. Boxplot of slope offset, *THEW tQT$_l$*-study, Moxifloxacin

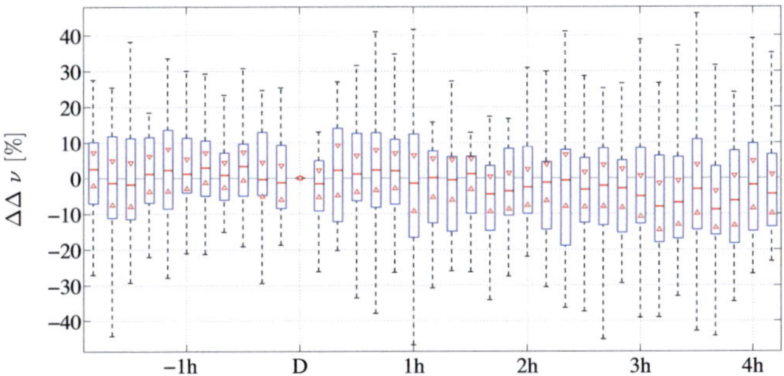

Fig. A.7. Boxplot of amplitude, *THEW tQT$_l$*-study, Moxifloxacin

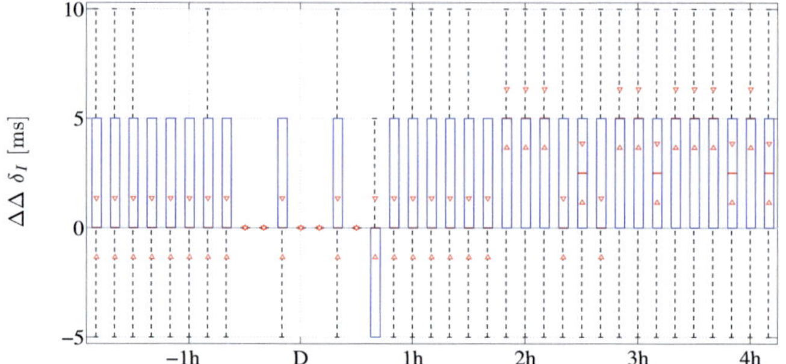

Fig. A.8. Boxplot of time$_I$ (T_{peak} to T_{end}, *THEW* tQT_I-study, Moxifloxacin)

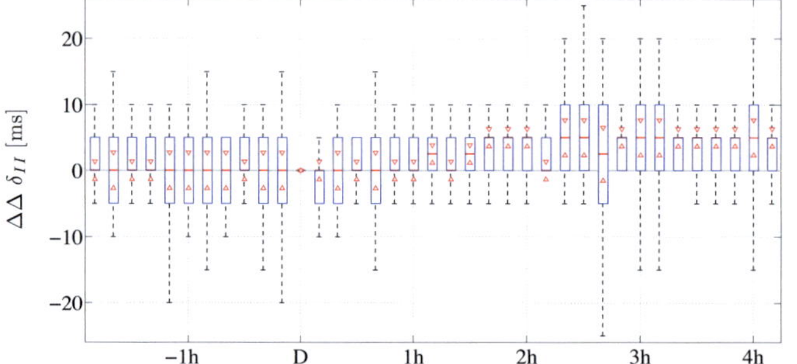

Fig. A.9. Boxplot of time$_{II}$ (T_{onset} to T_{end}, *THEW* tQT_I-study, Moxifloxacin)

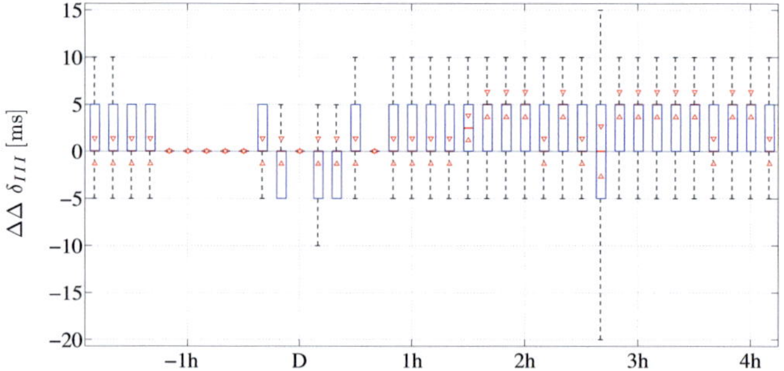

Fig. A.10. Boxplot of time$_{III}$ (T_{RP_I} to $T_{RP_{II}}$, *THEW* tQT_I-study, Moxifloxacin)

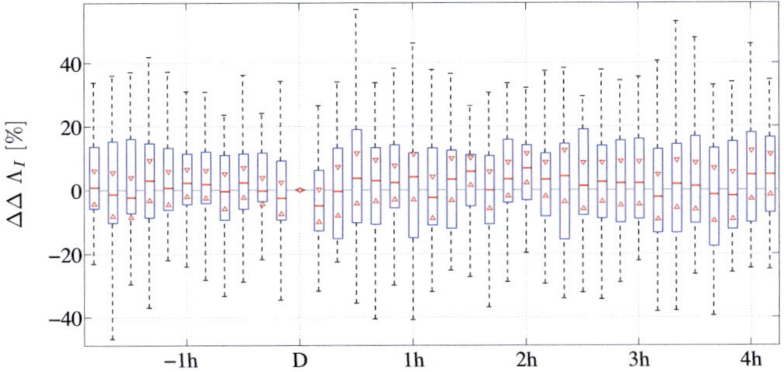

Fig. A.11. Boxplot of $area_I$ (T_{onset} to T_{peak}), *THEW tQT$_I$*-study, Moxifloxacin

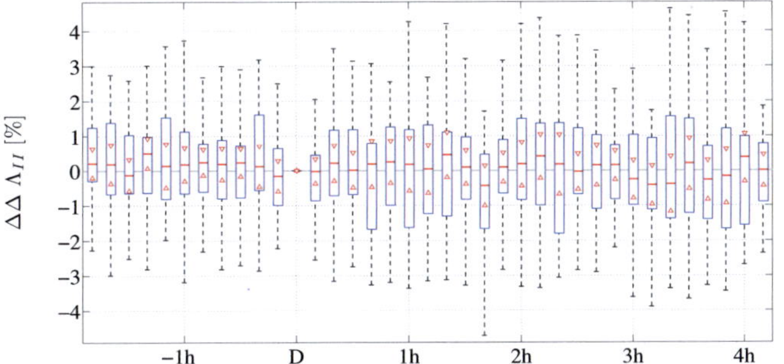

Fig. A.12. Boxplot of $area_{II}$ (T_{peak} to T_{end}), *THEW tQT$_I$*-study, Moxifloxacin

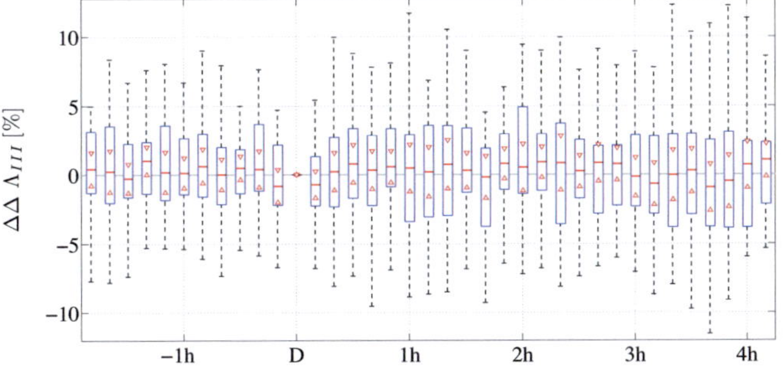

Fig. A.13. Boxplot of $area_{III}$ (T_{onset} to T_{end}), *THEW tQT$_I$*-study, Moxifloxacin

A.2.2 THEW tQT$_{II}$-Study Unknown Compound

Drug induced changes of T wave morphology depicting descriptors of 34 Subjects of *THEW tQT$_I$*-study. Interval length is 10 minutes. Time of dose is marked as D. The name of the compound is unknown in this study.

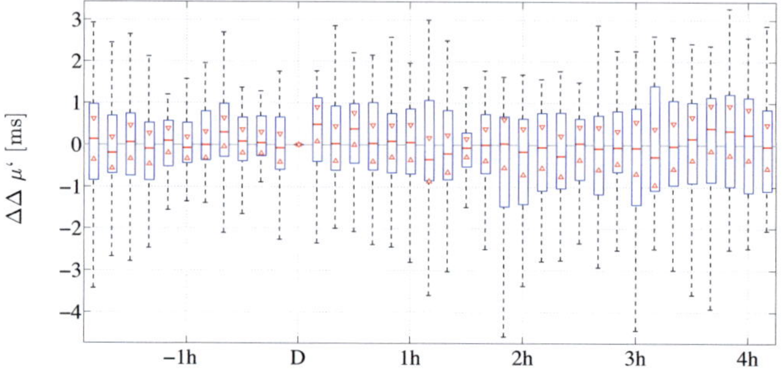

Fig. A.14. Boxplot of expected value, *THEW tQT$_I$*-study, unknown compound

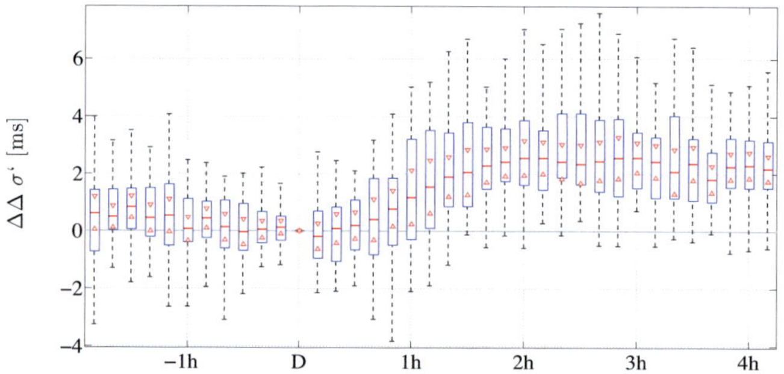

Fig. A.15. Standard Deviation, *THEW tQT$_I$*-study, unknown compound

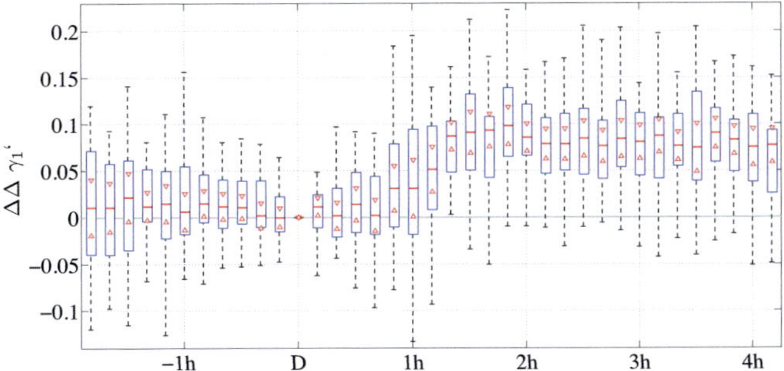

Fig. A.16. Boxplot of skewness, *THEW tQT$_l$*-study, unknown compound

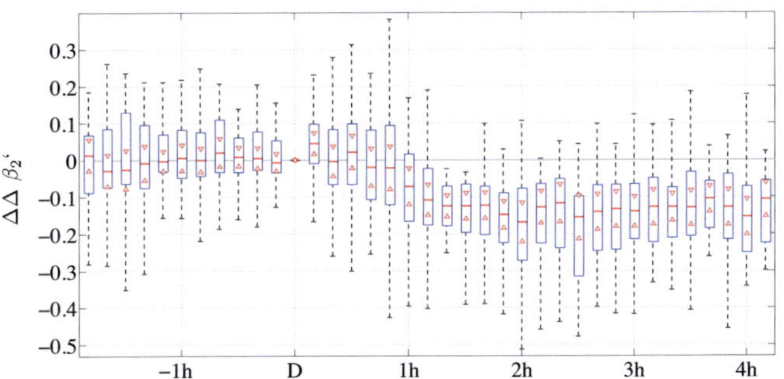

Fig. A.17. Boxplot of kurtosis, *THEW tQT$_l$*-study, unknown compound

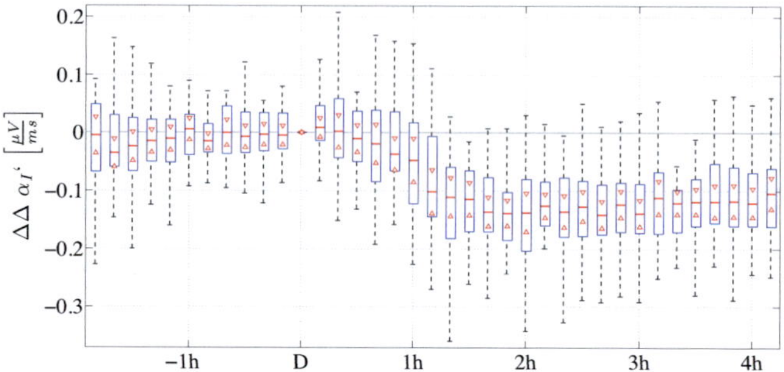

Fig. A.18. Boxplot of slope onset, *THEW tQT$_I$*-study, unknown compound

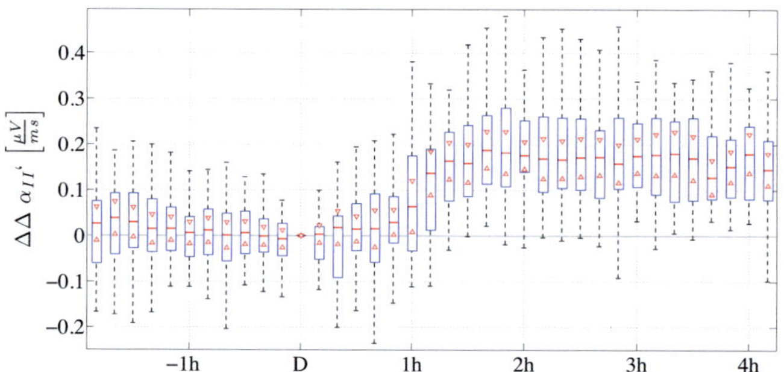

Fig. A.19. Boxplot of slope offset, *THEW tQT$_I$*-study, unknown compound

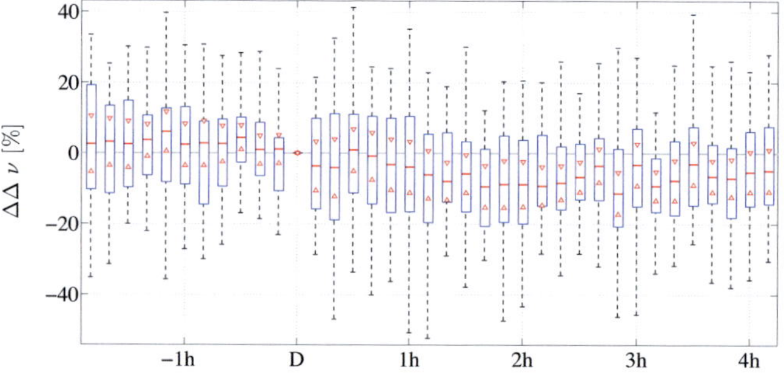

Fig. A.20. Boxplot of amplitude, *THEW tQT$_I$*-study, unknown compound

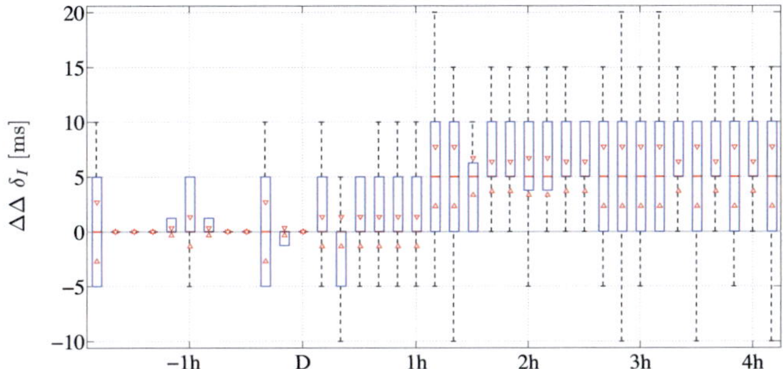

Fig. A.21. Boxplot of time$_I$ (T_{peak} to T_{end}, *THEW tQT$_I$*-study,unknown compound)

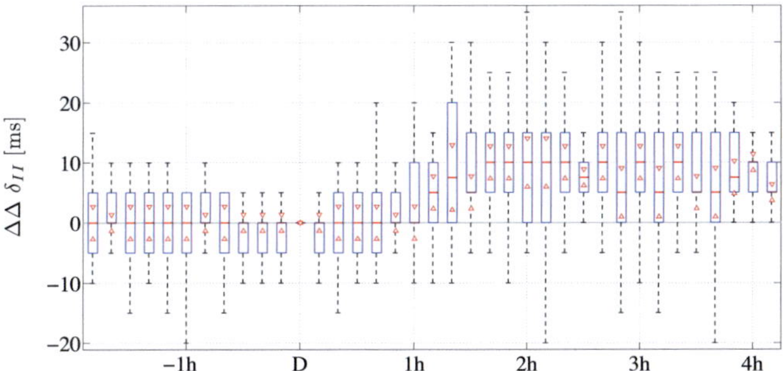

Fig. A.22. Boxplot of time$_{II}$ (T_{onset} to T_{end}), *THEW tQT$_I$*-study, unknown compound

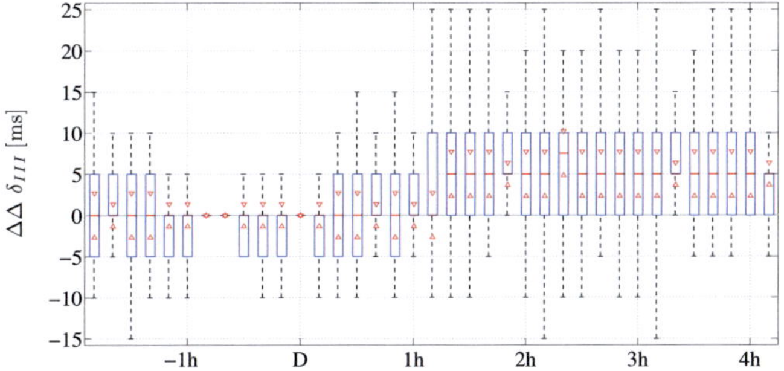

Fig. A.23. Boxplot of time$_{III}$ (T_{RP_I} to $T_{RP_{II}}$), *THEW tQT$_I$*-study, unknown compound

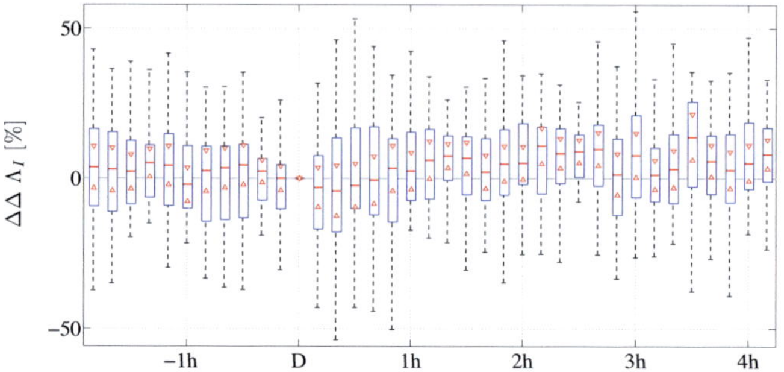

Fig. A.24. Boxplot of area$_I$ (T_{onset} to T_{peak}), *THEW tQT$_I$*-study, unknown compound

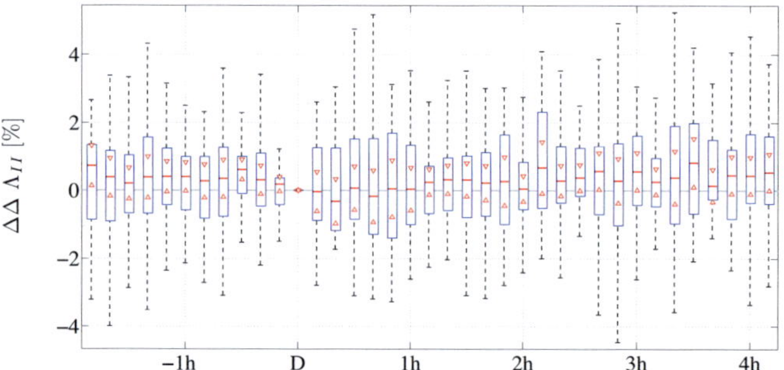

Fig. A.25. Boxplot of area$_{II}$ (T_{peak} to T_{end}), *THEW tQT$_I$*-study, unknown compound

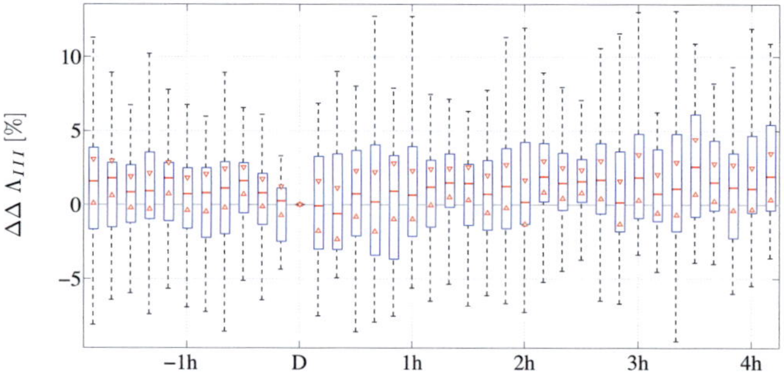

Fig. A.26. Boxplot of area$_{III}$ (T_{onset} to T_{end}), *THEW tQT$_I$*-study, unknown compound

References

1. Y. G. Yap and A. J. Camm, "Drug induced QT prolongation and torsades de pointes," *Heart (British Cardiac Society)*, vol. 89, pp. 1363–1372, 2003.

2. F. Gravenhorst, *New Approach for T-Wave Detection for Improved QT Studies.* Diploma Thesis, Karlsruhe: Institut für Biomedizinische Technik, Karlsruher Institut für Technologie (KIT), 2010.

3. T. Baas, F. Gravenhorst, H. Medhat, and et. al., "Detecting end of T-wave in ECG using a correlation based method," in *Proceedings of Biosignal*, (Berlin, 14.-16. July), 2010.

4. G. Lenis Parra, *Automatic detection and classification of ectopic beats in the ECG using a support vector machine.* Student Research Project, Karlsruhe: Institut für Biomedizinische Technik, Karlsruher Institut für Technologie (KIT), 2010.

5. T. Baas, F. Gravenhorst, R. Fischer, and et. al., "Comparison of three t-wave delineation algorithms based on wavelet filterbank, correlation and PCA," in *Computing in Cardiology*, vol. 37, (Belfast, 26.-29.Sept.), pp. 361–364, 2010.

6. T. Baas, K. Gräfe, A. Khawaja, and et. al., "Investigation of parameters highlighting drug induced small changes of the T-wave," in *Conf Proc IEEE Eng Med Biol Soc*, pp. 3796–3799, 2011.

7. K. Gräfe, *Untersuchung des Einflusses von Moxifloxacin auf die Morphologie des EKG-Signals zur Verbesserung der QT-Analyse.* Master Thesis, Karlsruhe: Institut für Biomedizinische Technik, Karlsruher Institut für Technologie (KIT), 2011.

8. M. Pfeifer, *Investigation of the heart rate's influence to the QT-RR dynamics and the morphology of the T-wave in healthy subjects.* Bachelor Thesis, Karlsruhe: Institute of Biomedical Engineering, Karlsruhe Institute of Technology (KIT), 2012.

9. H. Köhler, *Entwicklung eines multivariaten autoregressiven Modells zur Identifizierung von Myokarditis Erkrankungen.* Diploma Thesis, Karlsruhe: Institut für Biomedizinische Technik, Karlsruher Institut für Technologie (KIT), 2010.

10. T. Oesterlein, *Multivariate AR model parameter estimation on time series extracted from the ECG and BP of myocarditis patients.* Student Research Project, Karlsruhe: Institute of Biomedical Engineering, Karlsruhe Institute of Technology (KIT), 2011.

11. T. Oesterlein, T. Baas, H. Malberg, and et. al., "Multivariate AR model parameter estimation on time series extracted from the ECG of myocarditis patients," in *Biomedizinische Technik / Biomedical Engineering (Proceedings BMT2011)*, vol. 56, 2011.

12. G. Lenis, *Analysing rhythmical and morphological ECG properties to detect the influence of ectopic beats.* Diploma Thesis, Karlsruhe: Institut für Biomedizinische Technik, Karlsruher Institut für Tech-

nologie (KIT), 2012.

13. F. H. Martini, *Fundamentals of Anatomy & Physiology*. Prentice Hall, Inc. Upper Saddle River, NJ: Benjamin-Cummings Publishing Company, 2006.

14. K. S. Saladin, *Anatomy and Physiology: The Unity & Form of Function*. Mc Graw-Hill, 2004.

15. F. H. Netter, *Atlas der Anatomie des Menschen*. Stuttgart; New York: Thieme, 1997.

16. P. A. H. Iaizzo, ed., *Handbook of cardiac anatomy, physiology, and devices*. New York: Springer, 2. ed. ed., 2009.

17. R. Klinke and S. Silbernagl, *Taschenatlas der Physiologie*. Stuttgart; New York: Georg Thieme Verlag, 2001.

18. Z. Fari, "Heart: anterior view, coronal section." Loaded: Apr. 2012, `http://en.wikipedia.org/wiki/Human_heart`, License: [CC-BY-SA-3.0 (`http://creativecommons.org/licenses/by-sa/3.0`)].

19. G. Seemann, *Modeling of electrophysiology and tension development in the human heart*. PhD thesis, Universitätsverlag Karlsruhe, Institut für Biomedizinische Technik, 2005.

20. R. F. Schmidt, *Physiologie kompakt*. Springer-Lehrbuch, Berlin: Springer, 3., komplett überarb. aufl. ed., 1999.

21. J. Heuser, "Heart; conduction system." Loaded: Apr. 2012, `http://en.wikipedia.org/wiki/Electrical_conduction_system_of_the_heart`, License: [CC-BY-2.5 (`http://creativecommons.org/licenses/by/2.5`)].

22. H.-J. Schuster, Hans-Peter ; Trappe, *EKG-Kurs für Isabel*. Stuttgart: Thieme, 3., erw. aufl. ed., 2001.

23. J. W. Mason, D. J. Ramseth, D. O. Chanter, and et. al., "Electrocardiographic reference ranges derived from 79,743 ambulatory subjects," *Journal of Electrocardiology*, vol. 40, pp. 228–234, 2007.

24. B. Hopenfeld and H. Ashikaga, "Origin of the electrocardiographic U wave: effects of M cells and dynamic gap junction coupling," *Annals of Biomedical Engineering*, vol. 38, pp. 1060–1070, 2010.

25. H. J. Ritsema van Eck, J. A. Kors, and G. van Herpen, "The U wave in the electrocardiogram: a solution for a 100-year-old riddle," *Cardiovascular Research*, vol. 67, pp. 256–262, 2005.

26. B. Surawicz, "U.wave: facts, hypotheses, misconceptions, and misnomers," *Journal of Cardiovascular Electrophysiology*, vol. 9, pp. 1117–1128, 1998.

27. W. Einthoven, G. Fahr, and A. de Waart, "Über die Richtung und die manifeste Grösse der Potentialschwankungen im menschlichen Herzen und über den Einfluss der Herzlage auf die Form des Elektrokardiogramms," *Pflügers Archiv European Journal of Physiology*, vol. 150, pp. 275–315, 1913.

28. E. Goldberger, "A simple, indifferent, electrocardiographic electrode of zero potential and a technique of obtaining augmented, unipolar, extremity leads," *American Heart Journal*, vol. 23, pp. 483–492, 1942.

29. F. N. Wilson, F. D. Johnston, A. G. Macleod, and et. al., "Electrocardiograms that represent the potential variations of a single electrode," *American Heart Journal*, vol. 9, pp. 447–458, 1934.

30. M. Gertsch, *The ECG : a two-step approach to diagnosis*. Berlin: Springer, 2004. Includes bibliographical references.

31. U. Kiencke and H. Jäkel, *Signale und Systeme*. München: Oldenbourg-Verl., 2008.

32. E. O. Brigham and R. E. Morrow, "The fast Fourier transform," *Spectrum, IEEE*, vol. 4, pp. 63 –70, dec. 1967.

33. U. Kiencke, M. Schwarz, and T. Weickert, *Signalverarbeitung : Zeit-Frequenz-Analyse und Schätzverfahren*. München: Oldenbourg, 2008.

34. A. H. Aldroubi, ed., *Wavelets in medicine and biology*. Boca Raton [u.a.]: CRC Press, 1996.

35. H. Hotelling, "Analysis of a complex of statistical variables into principal components," *J. Educ. Psych.*, vol. 24, 1933.

36. A. J. Izenman, *Modern multivariate statistical techniques: Regression, classification, and manifold learning*. Springer texts in statistics, New York: Springer, 2008.

37. A. Khawaja, *Automatic ECG analysis using principal component analysis and wavelet transformation*. PhD thesis, Karlsruhe, Universitaetsverlag, 2006.

38. A. Rasiah, R. Togneri, and Y. Attikiouzel, "Modelling 1-D signals using Hermite basis functions," *Vision, Image and Signal Processing, IEE Proceedings -*, vol. 144, pp. 345 –354, dec 1997.

39. L. Sornmo, P. O. Borjesson, M. E. Nygards, and et. al., "A method for evaluation of QRS shape features using a mathematical model for the ECG," *IEEE Transactions on bio-Medical Engineering*, vol. 28, pp. 713–717, 1981.

40. C. Weiß, *Basiswissen medizinische Statistik*. lehrbuch-medizin.de, Heidelberg: Springer, 5., überarb. aufl. ed., 2010.

41. E. T. Federighi, "Extended Tables of the Percentage Points of Student's t-Distribution," *Journal of the American Statistical Association*, vol. 54, no. 287, pp. pp. 683–688, 1959.

42. W. G. Snedecor, George W. ; Cochran, *Statistical methods*. Ames, Iowa: Iowa State Univ. Press, 6. ed. ed., 1967.

43. J. D. H. Bronzino, ed., *The biomedical engineering handbook*. A CRC handbook, Boca Raton, Fla.: CRC Press, 1995.

44. *Electric circuits*. Upper Saddle River, NJ: Pearson Prentice Hall, 7. ed. ed., 2005.

45. Analog Devices, Inc., *JFET Input Instrumentation Amplifier with Rail-to-Rail Output in MSOP Package AD8220, revision B*, 2010. Loaded: May 2011, `http://www.atmel.com/dyn/resources/prod_documents/doc8067.pdf`.

46. Maxim Integrated Products, *3.3V/5V/Adjustable-Output, Step-Up DC-DC Converters, revision 2*, 1995. Loaded: May 2011, `http://datasheets.maxim-ic.com/en/ds/MAX756-MAX757.pdf`.

47. Atmel Corporation, *ATxmega64A1/128A1/192A1/256A1/384A1 Preliminary, revision M*, Sept 2010. Loaded: May 2011, `http://www.atmel.com/dyn/resources/prod_documents/doc8067.pdf`.

48. A. Dick, "Alvidi-business." `http://www.alvidi.de`.

49. J. Ottenbacher and M. Kirst, "Unisens 2.0." Loaded: Apr. 2010, `http://www.unisens.org/downloads/documentation/Unisens-Dokumentation.pdf`.

50. G. B. Moody, R. G. Mark, and A. L. Goldberger, "PhysioNet: a Web-based resource for the study of physiologic signals," *Engineering in Medicine and Biology Magazine, IEEE*, vol. 20, pp. 70–75, 2001.

51. P. Laguna, R. G. Mark, A. Goldberg, and et. al., "A database for evaluation of algorithms for measurement of QT and other waveform intervals in the ECG," in *Computers in Cardiology 1997*, pp. 673–676, 1997.

52. J. Couderc, "The telemetric and holter ECG warehouse initiative (THEW): A data repository for the design, implementation and validation of ECG-related technologies," in *Engineering in Medicine and*

Biology Society (EMBC), 2010 Annual International Conference of the IEEE, pp. 6252–6255, 2010.

53. THEW, "Thorough qt study # 1." Loaded: Apr. 2010, `http://www.thew-project.org/Database/E-HOL-03-0102-005.html`.

54. THEW, "Thorough qt study # 2." Loaded: Okt. 2011, `http://thew-project.org/Database/E-HOL-12-0140-008.html`.

55. T. Baas, H. Köhler, H. Malberg, and et. al., "Automatic blood pressure segmentation algorithm for analysing morphology changes," in *Biomedizinische Technik / Biomedical Engineering (Proceedings BMT2010)*, vol. 55, pp. 168–171, 2010.

56. A. L. Goldberger, L. A. Amaral, L. Glass, and et. al., "PhysioBank, PhysioToolkit, and PhysioNet: components of a new research resource for complex physiologic signals," *Circulation*, vol. 101, pp. E215–20, 2000.

57. G. B. Moody and R. G. Mark, "The impact of the MIT-BIH arrhythmia database," *IEEE Engineering in Medicine and Biology Magazine : the Quarterly Magazine of the Engineering in Medicine & Biology Society*, vol. 20, pp. 45–50, 2001.

58. G. D. H. Clifford, ed., *Advanced methods and tools for ECG data analysis*. Artech House engineering in medicine and biology series, Boston, Mass.[u.a.]: Artech House, 2006.

59. L. Sörnmo and P. Laguna, *Bioelectrical signal processing in cardiac and neurological applications*. Academic Press series in biomedical engineering, Amsterdam: Elsevier Academic Press, 2006.

60. B. U. Köhler, C. Henning, and R. Orglmeister, "The Principles of Software QRS Detection," *IEEE Engineering in Medicine and Biology*, vol. 1, pp. 42–57, 2002.

61. J. Pan and W. J. Tompkins, "A Real-Time QRS Detection Algorithm," *IEEE Transactions on Biomedical Engineering*, vol. 32, pp. 230–236, 1985.

62. P. Laguna, R. Jane, and P. Caminal, "Automatic detection of wave boundaries in multilead ECG signals: validation with the CSE database," *Computers and Biomedical Research, an International Journal*, vol. 27, pp. 45–60, 1994.

63. S. Mallat and W. L. Hwang, "Singularity detection and processing with wavelets," *Information Theory, IEEE Transactions on*, vol. 38, pp. 617–643, 1992.

64. C. Li, C. Zheng, and C. Tai, "Detection of ECG characteristic points using wavelet transforms," *Biomedical Engineering, IEEE Transactions on*, vol. 42, pp. 21–28, 1995.

65. M. Bahoura, M. Hassani, and M. Hubin, "DSP implementation of wavelet transform for real time ECG wave forms detection and heart rate analysis," *Computer Methods and Programs in Biomedicine*, vol. 52, pp. 35–44, 1997.

66. J. S. Sahambi, S. N. Tandon, and R. K. P. Bhatt, "Using wavelet transforms for ECG characterization. An on-line digital signal processing system," *Engineering in Medicine and Biology Magazine, IEEE*, vol. 16, pp. 77–83, 1997.

67. J. F. Bohnert, A. Khawaja, and O. Doessel, "ECG segmentation using wavelet transformation," in *41. Jahrestagung der DGBMT im VDE. Proceedings BMT 2007*, vol. 52, 2007.

68. J. P. Martinez, R. Almeida, S. Olmos, and et. al., "A wavelet-based ECG delineator: evaluation on standard databases," *Biomedical Engineering, IEEE Transactions on*, vol. 51, pp. 570–581, 2004.

69. E. Lepeschkin and B. Surawicz, "The measurement of the Q-T interval of the electrocardiogram," *Circulation*, vol. 6, pp. 378–388, 1952.

70. Q. Zhang, A. I. Manriquez, C. Medigue, and et. al., "An algorithm for robust and efficient location of T-wave ends in electrocardiograms," *IEEE Transactions on bio-Medical Engineering*, vol. 53,

pp. 2544–2552, 2006.

71. V. Krasteva, I. Jekova, and I. Christov, "Automatic detection of premature atrial contractions in the electrocardiogram," *Electrotechniques & Electronics E+E*, vol. 9-10, pp. 49–55, 2006.

72. J. Millet, M. A. Perez, G. Joseph, and et. al., "Previous identification of QRS onset and offset is not essential for classifying QRS complexes in a single lead," in *Computers in Cardiology 1997* (Ieee, ed.), pp. 299–302, 1997.

73. C. Maier, H. Dickhaus, and J. Gittinger, "Unsupervised morphological classification of QRS complexes," in *Computers in Cardiology, 1999*, pp. 683–686, 1999.

74. U. R. Acharya, J. Suri, J. A. E. Spann, and K. S.M., eds., *Advances in cardiac signal processing*. Berlin: Springer, 2007.

75. L. Di Marco and L. Chiari, "A wavelet-based ECG delineation algorithm for 32-bit integer online processing," *Biomedical Engineering Online*, vol. 10, p. 23, 2011.

76. Fda, "The clinical evaluation of QT/QTc interval prolongation and proarrhythmic potential for non-antiarrhythmic drugs," vol. ICH E14 Step 2, pp. 1–5, 2004.

77. MathWorks, *Signal Processing Toolbox TM Users Guide*. Natick, MA: The MathWorks, Inc., 2012.

78. A. Khawaja, R. Petrovic, A. Safer, and et. al., "Analyzing Thorough QT Study 1 & 2 in the Telemetric and Holter ECG Warehouse (THEW) using Hannover ECG System HES: A validation study," in *Computing in Cardiology*, vol. 37, pp. 349–352, 2010.

79. P. Laguna, R. Mark, A. Goldberg, and G. Moody, "A database for evaluation of algorithms for measurement of QT and other waveform intervals in the ECG," in *Computers in Cardiology 1997*, pp. 673–676, sep 1997.

80. H. C. Bazett, "An analysis of time-relations of electrocardiograms," *Heart*, vol. 7, pp. 353–370, 1920.

81. B. Darpo, A. A. Fossa, J.-P. Couderc, and et. al., "Improving the precision of QT measurements," *Cardiology Journal*, vol. 18, pp. 401–410, 2011.

82. J. P. Martinez and S. Olmos, "Methodological principles of T wave alternans analysis: a unified framework," *IEEE Transactions on bio-Medical Engineering*, vol. 52, pp. 599–613, 2005.

83. D. R. Adam, J. M. Smith, S. Akselrod, and et. al., "Fluctuations in T-wave morphology and susceptibility to ventricular fibrillation," *Journal of Electrocardiology*, vol. 17, pp. 209–218, 1984.

84. C.-C. Chang and C.-J. Lin, "LIBSVM: A library for support vector machines," *ACM Transactions on Intelligent Systems and Technology*, vol. 2, pp. 27:1–27:27, 2011. Software available at http://www.csie.ntu.edu.tw/~cjlin/libsvm.

85. R. Almeida, S. Gouveia, A. P. Rocha, and et. al., "QT variability and HRV interactions in ECG: quantification and reliability," *IEEE Transactions on bio-Medical Engineering*, vol. 53, pp. 1317–1329, 2006.

86. L. S. Fridericia, "The duration of systole in an electrocardiogram in normal humans and in patients with heart disease. 1920," *Annals of Noninvasive Electrocardiology : the Official Journal of the International Society for Holter and Noninvasive Electrocardiology, Inc*, vol. 8, pp. 343–351, 2003.

87. F. De Ponti, E. Poluzzi, and N. Montanaro, "QT-interval prolongation by non-cardiac drugs: lessons to be learned from recent experience," *European Journal of Clinical Pharmacology*, vol. 56, pp. 1–18, 2000.

88. B. Fermini and A. A. Fossa, "The impact of drug-induced QT interval prolongation on drug discovery and development," *Nature Reviews. Drug Discovery*, vol. 2, pp. 439–447, 2003.

89. R. R. Shah, "The significance of QT interval in drug development," *British Journal of Clinical Pharmacology*, vol. 54, pp. 188–202, 2002.

90. M. Malik, "Problems of heart rate correction in assessment of drug-induced QT interval prolongation," *Journal of Cardiovascular Electrophysiology*, vol. 12, pp. 411–420, 2001.

91. U. S. D. of Food and D. Administration, "International Conference on Harmonisation; guidance on E14 Clinical Evaluation of QT/QTc Interval Prolongation and Proarrhythmic Potential for Non-Antiarrhythmic Drugs; availability. Notice," *Federal Register*, vol. 70, pp. 61134–61135, 2005.

92. M. M. Hutmacher, S. Chapel, M. A. Agin, and et. al., "Performance characteristics for some typical QT study designs under the ICH E-14 guidance," *Journal of Clinical Pharmacology*, vol. 48, pp. 215–224, 2008.

93. C. Antzelevitch, "Ionic, molecular, and cellular bases of QT-interval prolongation and torsade de pointes," *Europace : European Pacing, Arrhythmias, and Cardiac Electrophysiology : Journal of the Working Groups on Cardiac Pacing, Arrhythmias, and Cardiac Cellular Electrophysiology of the European Society of Cardiology*, vol. 9 Suppl 4, pp. iv4–15, 2007.

94. M. Hodges, "Rate Correction of the QT Interval," *Cardiac Electrophysiology Review*, vol. 1, pp. 360–363, 1997. 10.1023/A:1009933509868.

95. E. Pueyo, P. Smetana, P. Laguna, and et. al., "Estimation of the QT/RR hysteresis lag," *Journal of Electrocardiology*, vol. 36 Suppl, pp. 187–190, 2003.

96. M. Tabo, M. Nakamura, K. Kimura, and et. al., "QT-RR relationships and suitable QT correction formulas for halothane-anesthetized dogs," *The Journal of Toxicological Sciences*, vol. 31, pp. 381–390, 2006.

97. E. Pueyo, P. Smetana, P. Caminal, and et. al., "Characterization of QT interval adaptation to RR interval changes and its use as a risk-stratifier of arrhythmic mortality in amiodarone-treated survivors of acute myocardial infarction," *IEEE Transactions on bio-Medical Engineering*, vol. 51, pp. 1511–1520, 2004.

98. J. Halamek, P. Jurak, M. Villa, and et. al., "Dynamic coupling between heart rate and ventricular repolarisation," *Biomed Tech (Berl)*, vol. 52, pp. 255–263, 2007.

99. T. Baas, G. Butrous, and D. O. Azulay, "Modeling the dynamics of QT-RR relationship," in *Tagungsband Biosignalverarbeitung*, vol. Potsdam, pp. 67–70, 2008.

100. T. Baas, G. Boutrous, and O. Dössel, "A cell-based approach in dynamic QT-Correction," in *IFMBE Proceedings, WC 2009* (O. Dössel and W. Schlegel, eds.), pp. 2144–2146, Springer, 2009.

101. T. Baas, *Modeling the Dynamics of QT-RR Relationship*. Diploma Thesis, Karlsruhe: Institute of Biomedical Engineering, Universität Karlsruhe (TH), 2008.

102. M. Malik, P. Farbom, V. Batchvarov, and et. al., "Relation between QT and RR intervals is highly individual among healthy subjects: implications for heart rate correction of the QT interval," *Heart*, vol. 87, pp. 220–228, 2002.

103. D. U. J. Keller, D. L. Weiss, O. Dössel, and et. al., "Influence of I(Ks) heterogeneities on the genesis of the T-wave: a computational evaluation," *IEEE Trans. Biomed. Engineering*, vol. 59, pp. 311–322, 2012.

104. D. U. J. Keller, *Multiscale Modeling of the Ventricles : IFrom Cellular Electrophysiology to Body Surface Electrocardiograms*. PhD thesis, Karlsruhe, KIT Scientific Publishing, 2011.

105. J.-P. Couderc, "Measurement and regulation of cardiac ventricular repolarization: from the QT interval to repolarization morphology," *Philosophical Transactions. Series A, Mathematical, Physical,*

and *Engineering Sciences*, vol. 367, pp. 1283–1299, 2009.

106. J.-P. Couderc, S. McNitt, O. Hyrien, and et. al., "Improving the detection of subtle I(Kr)-inhibition: assessing electrocardiographic abnormalities of repolarization induced by moxifloxacin," *Drug Safety : an International Journal of Medical Toxicology and Drug Experience*, vol. 31, pp. 249–260, 2008.

107. G. Malfatto, M. Facchini, and A. Zaza, "Characterization of the non-linear rate-dependency of QT interval in humans," *Europace : European Pacing, Arrhythmias, and Cardiac Electrophysiology : Journal of the Working Groups on Cardiac Pacing, Arrhythmias, and Cardiac Cellular Electrophysiology of the European Society of Cardiology*, vol. 5, pp. 163–170, 2003.

108. M. Malik, K. Hnatkova, A. Schmidt, and P. Smetana, "Correction for QT/RR hysteresis in the assessment of drug-induced QTc changes–cardiac safety of gadobutrol," *Annals of Noninvasive Electrocardiology*, vol. 14, no. 3, pp. 242–250, 2009.

109. H. Morita, "How can we stabilize QT variability?," *Heart Rhythm : the Official Journal of the Heart Rhythm Society*, vol. 8, pp. 1243–1244, 2011.

110. B. R. Chaitman, M. J. Moinuddin, and J. Sano, "Exercise Testing," in *Cardiovascular Medicine* (J. T. Willerson, H. J. J. Wellens, J. N. Cohn, and D. R. Holmes, eds.), pp. 729–744, Springer London, 2007. 10.1007/978-1-84628-715-24.

111. J.-P. Couderc, M. Vaglio, X. Xia, and et. al., "Impaired T-amplitude adaptation to heart rate characterizes I(Kr) inhibition in the congenital and acquired forms of the long QT syndrome," *Journal of Cardiovascular Electrophysiology*, vol. 18, pp. 1299–1305, 2007.

112. M. P. Andersen, J. Q. Xue, C. Graff, and et. al., "New descriptors of T-wave morphology are independent of heart rate," *Journal of Electrocardiology*, vol. 41, pp. 557–561, 2008.

113. M. Malik, K. Hnatkova, and V. Batchvarov, "Differences between study-specific and subject-specific heart rate corrections of the QT interval in investigations of drug induced QTc prolongation," *Pacing and Clinical Electrophysiology : PACE*, vol. 27, pp. 791–800, 2004.

114. J. Halamek, P. Jurak, M. Novak, and et. al., "Dynamic QT/RR Coupling in Patients with pacemaker," in *Proceedings of the 29th Annual International Conference of the IEEE EMBS*, (Cite Internationale, Lyon, France), pp. 919–922, 2007.

115. E. Lehmann, Thomas Martin; Ammenwerth, ed., *Handbuch der medizinischen Informatik*. München: Hanser, 2002.

116. R. Schlittgen and B. Streitberg, *Zeitreihenanalyse*. Lehr- und Handbücher der Statistik, München: Oldenbourg, 9., unwesentlich veränd. aufl. ed., 2001.

117. M. Malik, K. Hnatkova, T. Novotny, and et. al., "Subject-specific profiles of QT/RR hysteresis," *American Journal of Physiology. Heart and Circulatory Physiology*, vol. 295, pp. H2356–63, 2008.

118. A. Bauer and G. Schmidt, "Heart rate turbulence," *Journal of Electrocardiology*, vol. 36 Suppl, pp. 89–93, 2003.

119. G. Schmidt, M. Malik, P. Barthel, and et. al., "Heart-rate turbulence after ventricular premature beats as a predictor of mortality after acute myocardial infarction," *The Lancet*, vol. 353, 1999.

120. A. Bauer, M. Malik, G. Schmidt, and et. al., "Heart rate turbulence: standards of measurement, physiological interpretation, and clinical use: International Society for Holter and Noninvasive Electrophysiology Consensus," *Journal of the American College of Cardiology*, vol. 52, pp. 1353–1365, 2008.

121. G. Schmidt, "H-r-t.org." http://www.h-r-t.org.

List of Publications and Supervised Theses

Conference Contributions

- **T. Baas**, G. Butrous, and D. O. Azulay, *'Modeling the dynamics of QT-RR relationship'*, In Tagungsband Biosignalverarbeitung, vol. Potsdam, pp. 67-70, 2008

- **T. Baas**, G. Boutrous, and O. Dössel, *'A cell-based approach in dynamic QT-Correction'*, In IFMBE Proceedings, WC 2009, pp. 2144-2146, 2009

- **T. Baas**, H. Köhler, H. Malberg and O. Dössel, *'Automatic blood pressure segmentation algorithm for analysing morphology changes'*, In Biomedizinische Technik / Biomedical Engineering (Proceedings BMT2010), vol. 55(1) , pp. 168-171, 2010

- **T. Baas**, F. Gravenhorst, R. Fischer, A. Khawaja and O. Dössel, *'Comparison of three t-wave delineation algorithms based on wavelet filterbank, correlation and PCA'*, In Computing in Cardiology, vol. 37, pp. 361-364, 2010

- A. Khawaja, R. Petrovic, A. Safer, **T. Baas**, O. Dössel and R. Fischer, *'Analyzing Thorough QT Study 1 & 2 in the Telemetric and Holter ECG Warehouse (THEW) using Hannover ECG System HES: A validation study'*, In Computing in Cardiology, vol. 37, pp. 349-352, 2010

- **T. Baas**, F. Gravenhorst, H. Medhat and O. Dössel. *'Detecting end of T-wave in ECG using a correlation based method'*, In Proceedings of Biosignal, 2010

- **T. Baas**, K. Gräfe, A. Khawaja and O. Dössel, *'Investigation of parameters highlighting drug induced small changes of the T-wave'*, In Conf Proc IEEE Eng Med Biol Soc, pp. 3796-3799, 2011

- T. Oesterlein, **T. Baas**, H. Malberg and O. Dössel, *'Multivariate AR model parameter estimation on time series extracted from the ECG of myocarditis patients'* In Biomedizinische Technik / Biomedical Engineering (Proceedings BMT2011), vol. 56(1) , 2011

- G. Lenis, **T. Baas** and O. Dössel, *'Automatic Detection and Classification of Ectopic Beats in the ECG using a Support Vector Machine'* In Biomedizinische Technik / Biomedical Engineering (Proceedings BMT2011), vol. 56(1) , 2011

- G. Lenis, **T. Baas** and O. Dössel, *'Artifact detection in the electrocardiogram to produce a robust classification of ectopic beats'*, Berlin, Innovative Verarbeitung bioelektrischer und biomagnetischer Signale (Proceedings BBS2012), 2012

Supervised Diploma Theses and Student Research Projects

- Tobias Heiß '*Multivariate Zeitreihenanalyse von Langzeit EKGs,*' Diploma Thesis, Institute of Biomedical Engineering, University of Karlsruhe (TH), 2008

- Franz Gravenhorst, '*New Approach for T-Wave Detection for Improved QT Studies,*' Diploma Thesis, Institute of Biomedical Engineering, Karlsruhe Institute of Technology (KIT), 2010

- Heidrun Köhler, '*Entwicklung eines multivariaten autoregressiven Modells zur Identifizierung von Myokarditis Erkrankungen,*' Diploma Thesis, Institute of Biomedical Engineering, Karlsruhe Institute of Technology (KIT), 2010

- Gustavo Lenis Parra, '*Automatic detection and classification of ectopic beats in the ECG using a support vector machine*'. Student Research Project, Institute of Biomedical Engineering, Karlsruhe Institute of Technology (KIT), 2010

- Ksenja Gräfe, '*Untersuchung des Einflusses von Moxifloxacin auf die Morphologie des EKG-Signals zur Verbesserung der QT-Analyse*'. Master Thesis, Institute of Biomedical Engineering, Karlsruhe Institute of Technology (KIT), 2011

- Tobias Oesterlein, '*Multivariate AR model parameter estimation on time series extracted from the ECG and BP of myocarditis patients*'. Student Research Project, Institute of Biomedical Engineering, Karlsruhe Institute of Technology (KIT), 2011

- Martin Pfeifer, *'Investigation of the heart rates influence to the QT-RR dynamics and the morphology of the T-wave in healthy subjects'*. Bachelor Thesis, Institute of Biomedical Engineering, Karlsruhe Institute of Technology (KIT), 2012

- Gustavo Lenis Parra, *'Analysing rhythmical and morphological ECG properties to detect the influence of ectopic beats'*. Diploma Thesis, Institute of Biomedical Engineering, Karlsruhe Institute of Technology (KIT), 2012

Other Contributions

- S. Wieland, N. Kikillus, **T. Baas**, M. Braecklein and A. Bolz, *'Screening device for identification of patients with paroxysmal atrial fibrillation to prevent ischemic strokes'*, IEEE Engineering in Medicine and Biology Society. Conference, vol. 2007, pp. 3693-3696, 2007

- **Tobias Baas**, *'Konzeption, Entwicklung und Durchführung eines automatischen Performance-tests für die Erkennung von Vorhofflimmern'*, Student Research Project, Institute of Biomedical Engineering, Karlsruhe Institute of Technology (KIT), 2007

- **Tobias Baas**, *'Modeling the Dynamics of QT-RR Relationship'*, Diploma Thesis, Institute of Biomedical Engineering, University of Karlsruhe (TH), 2008

Karlsruhe Transactions on Biomedical Engineering
(ISSN 1864-5933)

Karlsruhe Institute of Technology / Institute of Biomedical Engineering (Ed.)

**Die Bände sind unter www.ksp.kit.edu als PDF frei verfügbar oder
als Druckausgabe bestellbar.**

Karlsruhe Transactions on Biomedical Engineering
(ISSN 1864-5933)